NP121-1

NP121-1

The Modern Mathematics Series, Second Edition

MEETING
MATHEMATICS

Edwina Deans

Robert B. Kane

George H. McMeen

Robert A. Oesterle

AMERICAN BOOK COMPANY

EDWINA DEANS

Specialist in elementary mathematics education; formerly Assistant Editor of *The Arithmetic Teacher;* previously Assistant Professor of Education at the University of Cincinnati, Elementary School Supervisor in Arlington County, Virginia, and elementary school teacher.

ROBERT B. KANE

Professor of Mathematics and Education and Coordinator of Student Teaching in Mathematics at Purdue University; previously a mathematics teacher with experience in the mathematics of industry. Dr. Kane is a mathematics consultant to schools, active in in-service work with elementary school teachers, and has written numerous articles for professional journals.

GEORGE H. McMEEN

Associate Professor of Mathematics at California State Polytechnic College; previously Chairman and Professor of Mathematics at Newark State College, Union, New Jersey, and Administrative Consultant in Mathematics to the California State Department of Education. Dr. McMeen is a mathematics consultant to schools, is active in in-service training programs, and has written articles for professional journals.

ROBERT A. OESTERLE

Formerly Professor of Mathematics Education at Purdue University; previously Director of Student Teaching at Purdue University, teacher of mathematics at various grade levels, mathematics consultant, subject matter associate and reviewer of televised mathematics programs for elementary schools, and author of secondary school mathematics texts and articles for professional journals.

Text illustrations by Romer/Todd *and* Jean Simpson
Cover illustrations by Jean Simpson

CONTENTS

subtraction facts through sums and minuends of 9; vertical form for addition and subtraction; recognizing coins, money values; comparing amounts of money; column addition of three numbers; hand symbols for 5 and 10; comparing numbers; using 5; counting by 1's, 2's, 3's, 4's, 5's; number sequence through 10; reading and writing numerals through '10'; chapter review; practice; enrichment.

HOW MANY?

How many? How do you know?

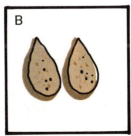

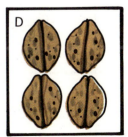

1

ONE FOR EACH

Find the first squirrel; the second squirrel; the last squirrel.

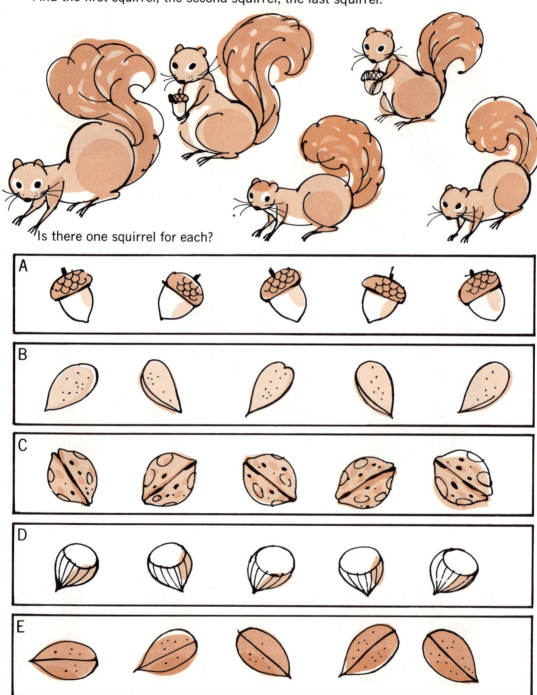

Is there one squirrel for each?

HOW MANY?

How many in each set? How do you know?

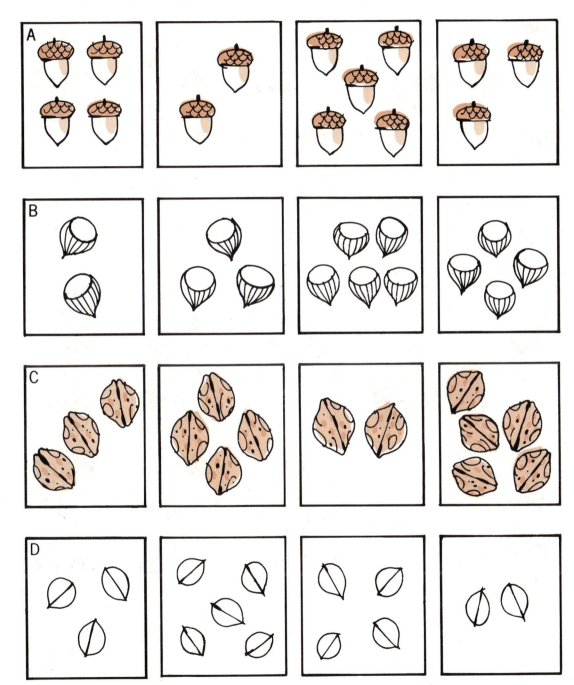

NUMERALS AND SETS

a. How many in each set?

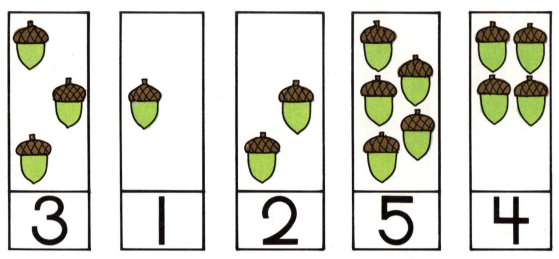

b. Match the sets with the numeral.

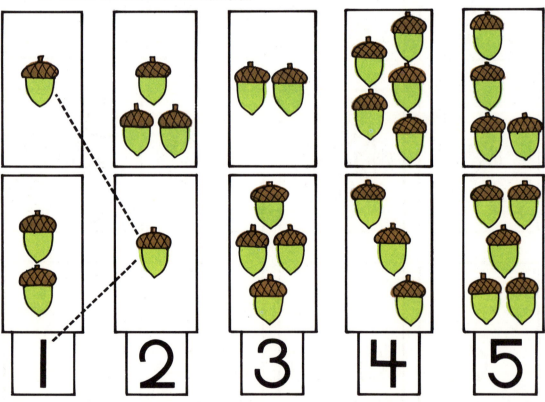

4

HOW MANY MORE?

1. The first set matches the numeral.

How many nuts does the second set need? the third? the fourth?

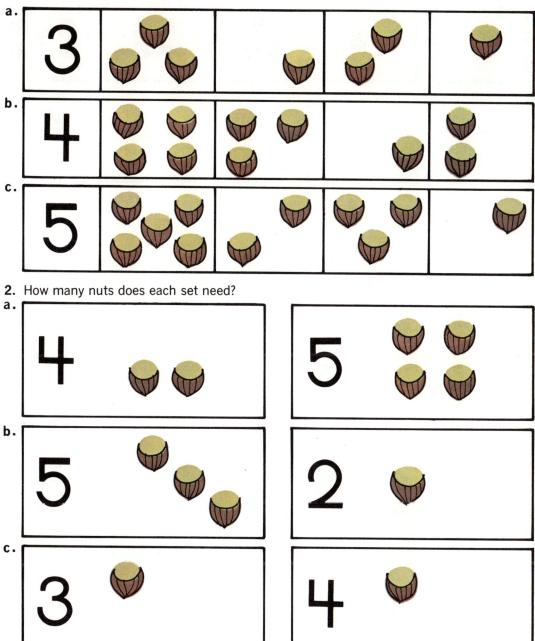

2. How many nuts does each set need?

READING AND WRITING NUMERALS

1. Read each numeral.

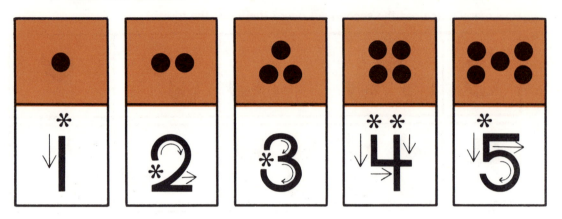

2. Which numeral is missing?

READING NUMBER WORDS

1. Read each number word.

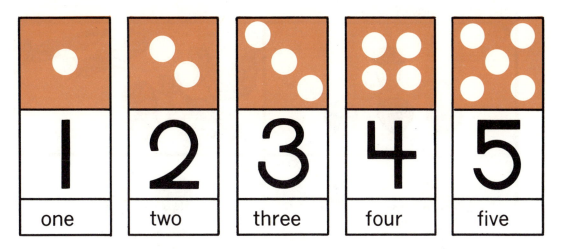

2. Match number words, sets, and numerals.

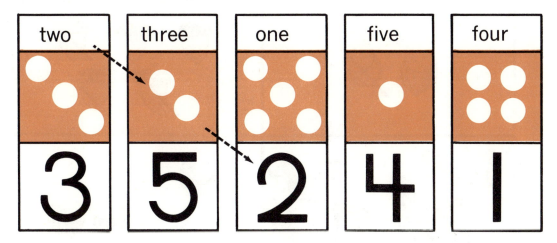

3. Write the numeral that tells how many are in the set.

PROBLEM SOLVING

Find a picture for each part of the story.

What is your answer?

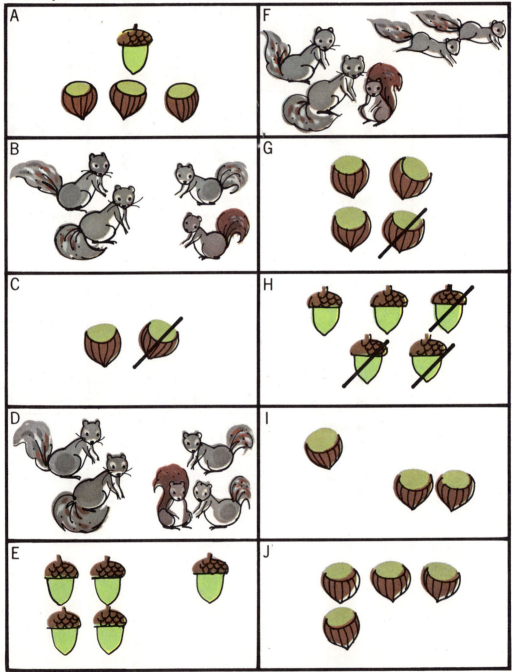

PROBLEM SOLVING

Tell a number story for each row.

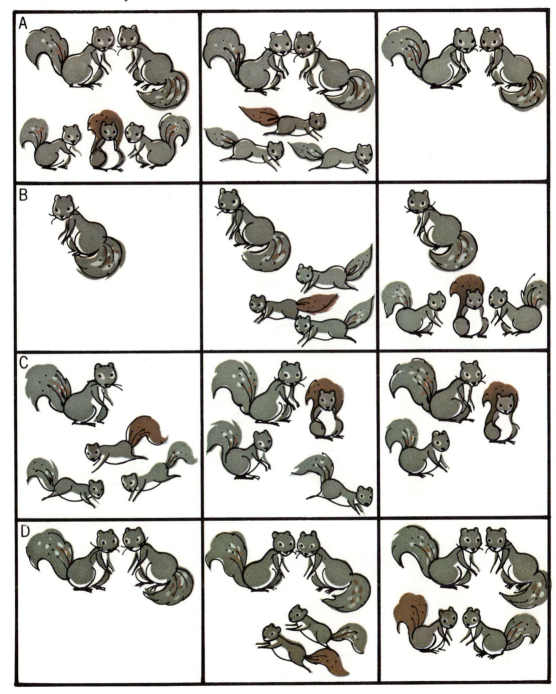

PRACTICE

1. Do you have enough beads? Tell how many more you need.

a.

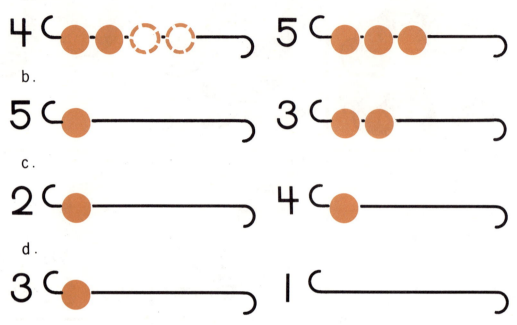

2. Do you have too many beads? Tell how many to take off.

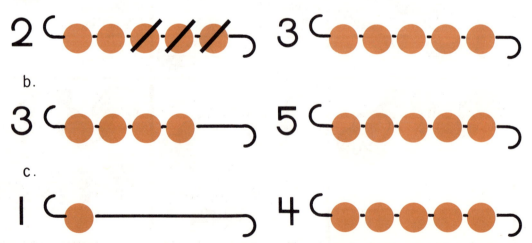

1. How many are there in each set?

a.

b.

c.

2. Match the numerals and sets of beads.

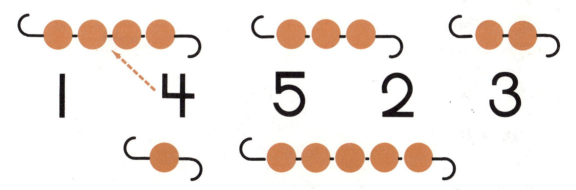

1   4   5   2   3

3. Draw the blocks you need to match the numerals.

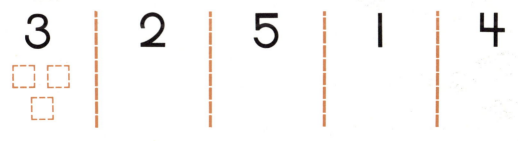

3   2   5   1   4

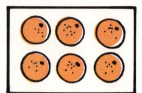

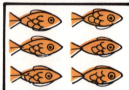

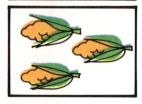

HOW MANY?

1. How many raccoons do you see?

2. How many are in each set?

ONE FOR EACH

How many rows?

How many are in each row?

What is in row 1? in row 3?

The acorns are in which row? Is there an acorn for each raccoon?

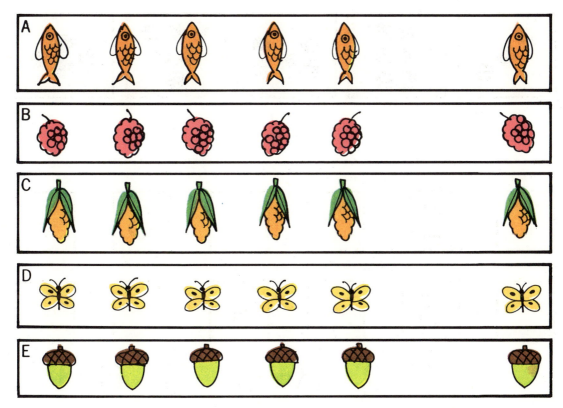

READING AND WRITING '6'

1. How many are in each set?

a.

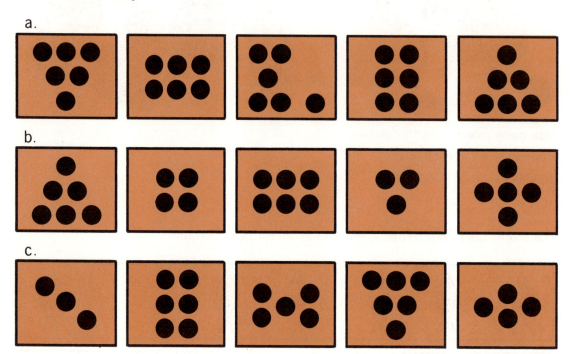

b.

c.

2. Find the missing numerals.

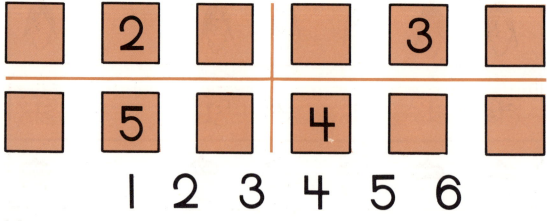

ADDING AND SUBTRACTING

1. How many are in each picture?

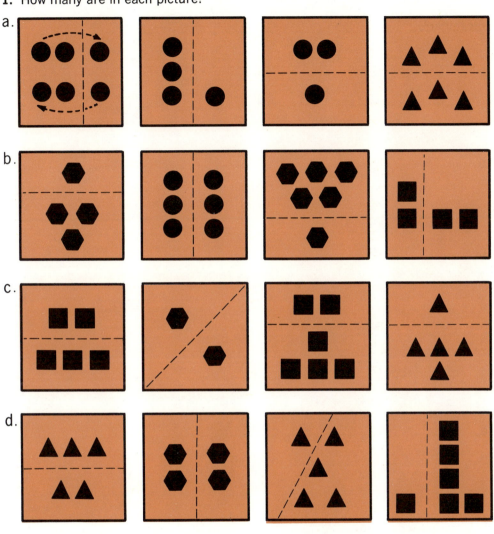

a.

b.

c.

d.

2. Cover some in each picture. How many do you have now?

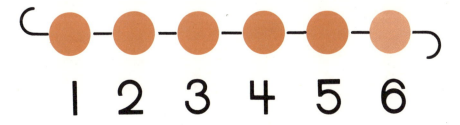

1   2   3   4   5   6

ADDING

How many are in each set? in both sets?

All together?

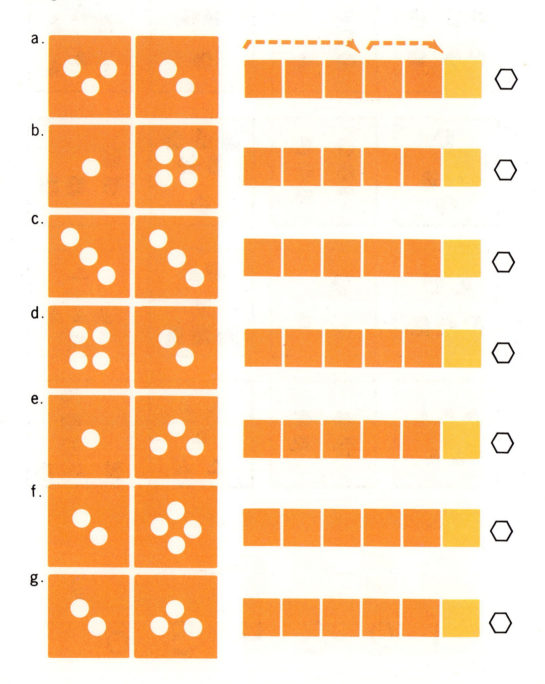

SUBTRACTING

How many in all are in the set? How many stay?

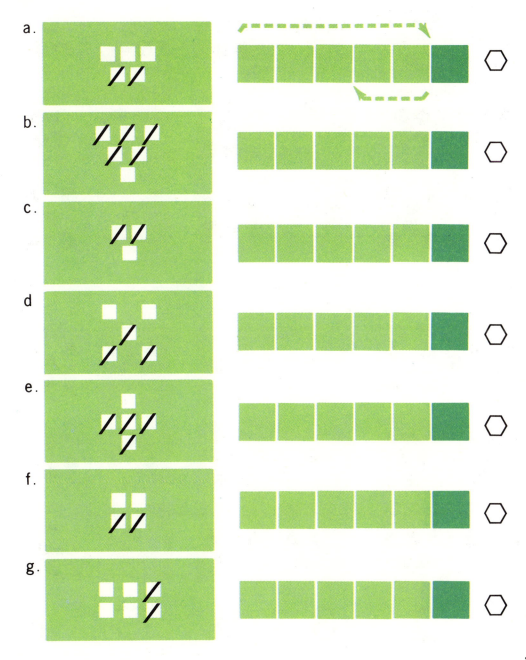

MORE AND LESS

1. How many in each set? Which set has more? How many more?

a   b   c

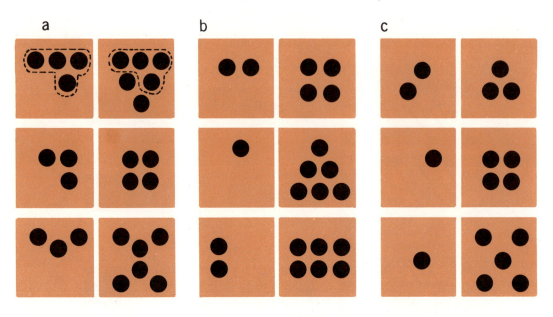

2. How many more do you need to make the sets match?

a   b   c

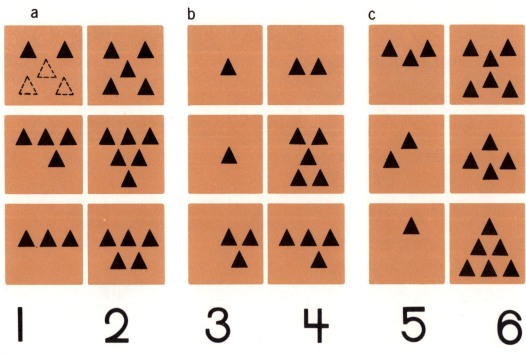

1   2   3   4   5   6

18

PROBLEM SOLVING

Find a set that fits each part of the story.

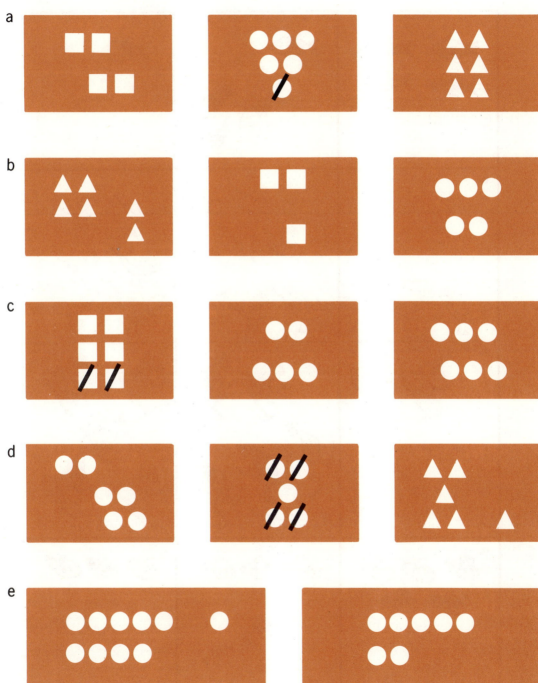

PROBLEM SOLVING

Tell a number story for each row.

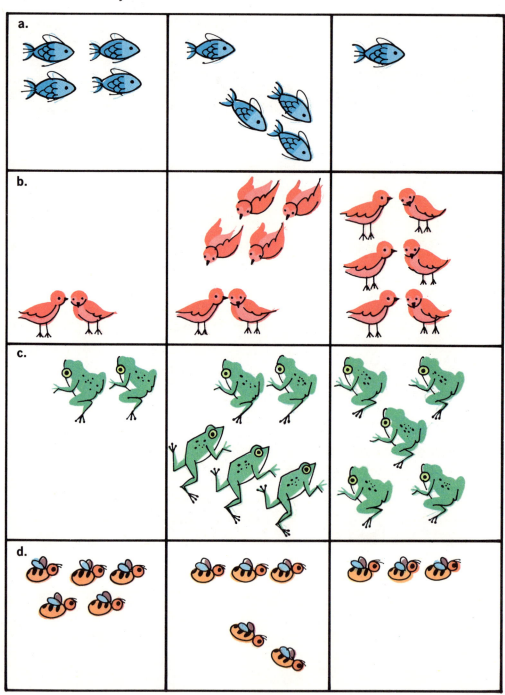

REVIEWING WHAT YOU KNOW

1. Which set has more? How many more?

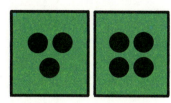

2. How many are in each set? in both sets?

A    B    C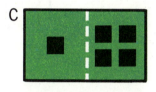

D    E    F

3. How many in all are in the set? How many stay?

A    B    C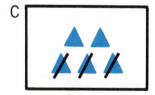

D   E 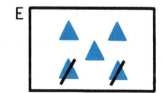   F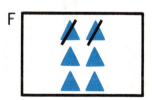

1 2 3 4 5 6

21

1. How many blocks are in each tower?

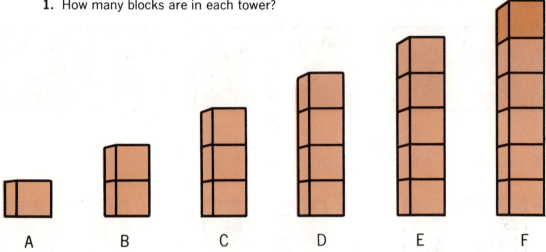

A   B   C   D   E   F

2. Count from F to A.

3. How many blocks are in each of these towers? B? D? F?

4. In each of these towers? A? C? E?

5. How many blocks are missing?

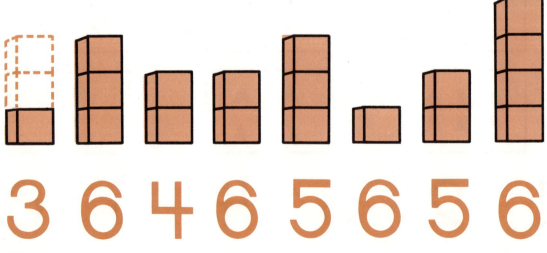

3 6 4 6 5 6 5 6

PRACTICE

How many blocks are in each tower?

Which has more? How many more?

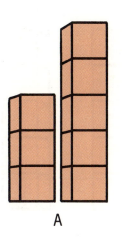

A

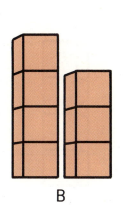

B

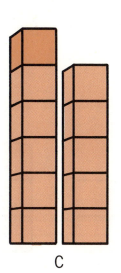

C

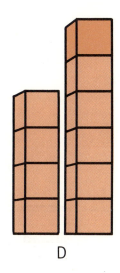

D

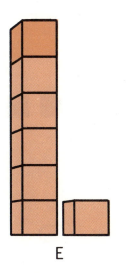

E

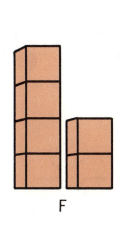

F

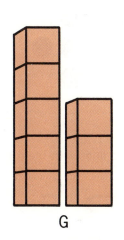

G

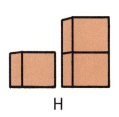

H

1 2 3 4 5 6

1. How many in B? D? F?

How many in A? C? E?

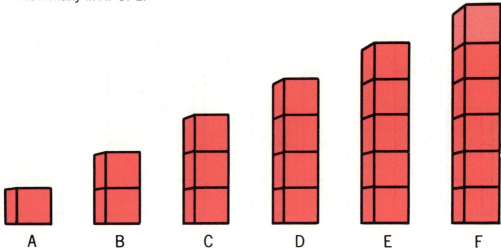

2. How many are in each set?

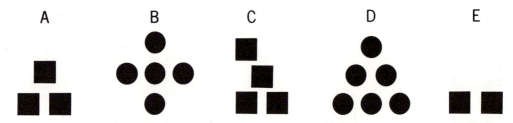

3. Match the bead lines with the numeral.

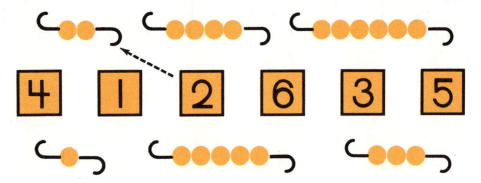

24

PRACTICE

1. Which numeral stands for the larger number? Match the numerals with the sets.

2. Read the numerals in A–D. Put them in order.

A	B	C	D
1	3	5	5
3	2	4	4
2	4	3	6

HOW MANY?

1 2 3 4 5 6 7

ONE FOR EACH

How many 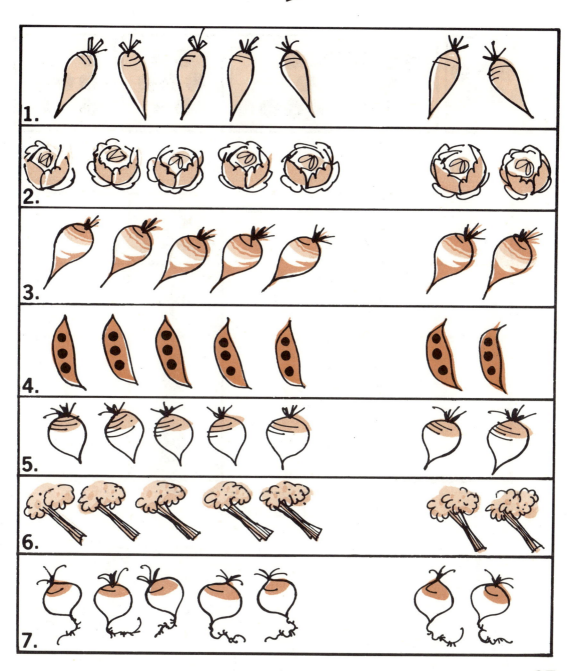's? Is there a for each ?

READING AND WRITING '7'

1. How many?

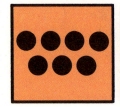 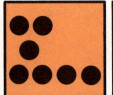 

2. Just before?   Just after?

	6
	4
	5
	7

2	
4	
6	
3	

3. What numerals are missing?

1		3
	5	

5	6	
3		5

28

TELLING HOW MANY

How many are in each set?
Which sets have the same number as the first set?

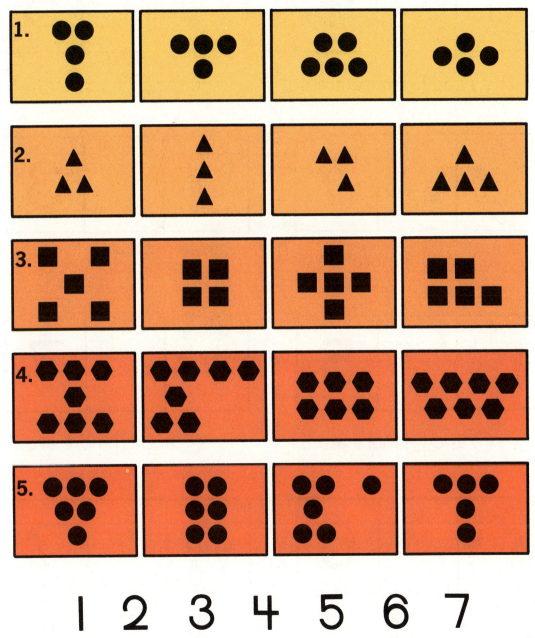

1 2 3 4 5 6 7

HOW MANY MORE?

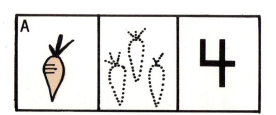

1 and 3 more are 4.

3 and 2 more are 5.

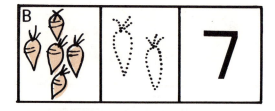

5 and ☐ more are 7.

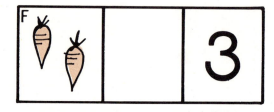

⬡ and ☐ more are 3.

⬡ and ☐ more are 5.

⬡ and ☐ more are 7.

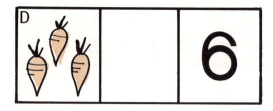

⬡ and ☐ more are 6.

⬡ and ☐ more are 7.

MORE OR LESS

1. How many more?

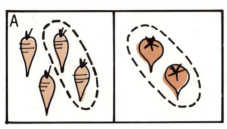

⟨4⟩ is ☐2 more than 2.

⬡ is ☐ more than 4.

⬡ is ☐ more than 3.

⬡ is ☐ more than 2.

2. How many less?

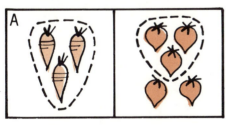

⟨3⟩ is ☐2 less than 5.

⬡ is ☐ less than 6.

⬡ is ☐ less than 6.

⬡ is ☐ less than 7.

31

USING SETS TO ADD

Tell a number story about each set.

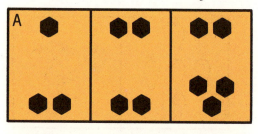 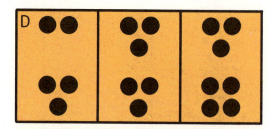

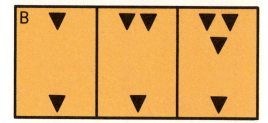 

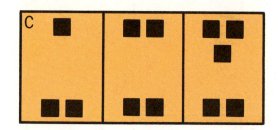

1. 2 add 2 = ☐   5 add 2 = ☐
3 add 2 = ☐   2 add 4 = ☐
6 add 1 = ☐   3 add 1 = ☐

2. 1 add 2 = ☐   2 add 3 = ☐
5 add 1 = ☐   1 add 4 = ☐
3 add 4 = ☐   3 add 3 = ☐

1 2 3 4 5 6 7

USING SETS TO SUBTRACT

Tell a number story about each set.

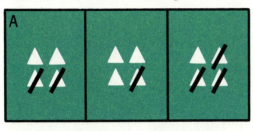 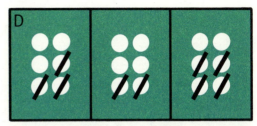

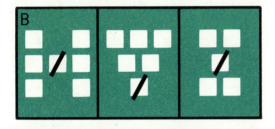 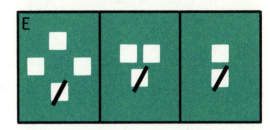

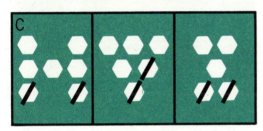 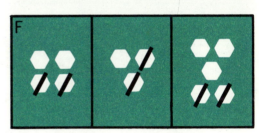

1. 3 subtract 1 = ☐   6 subtract 4 = ☐
    7 subtract 1 = ☐   4 subtract 3 = ☐
    6 subtract 2 = ☐   5 subtract 1 = ☐

2. 6 subtract 1 = ☐   4 subtract 2 = ☐
    4 subtract 1 = ☐   6 subtract 3 = ☐
    7 subtract 2 = ☐   3 subtract 2 = ☐

1   2   3   4   5   6   7

WHICH ONE IS IT?

1. Which  is eating?

2. Which is in the tree?

3. How many 's in each set?

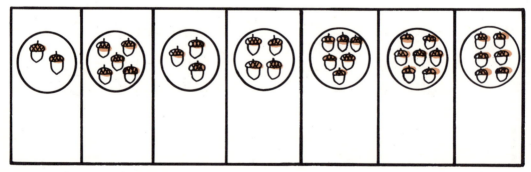

34

PROBLEM SOLVING

What did Hopper see? How many of each?

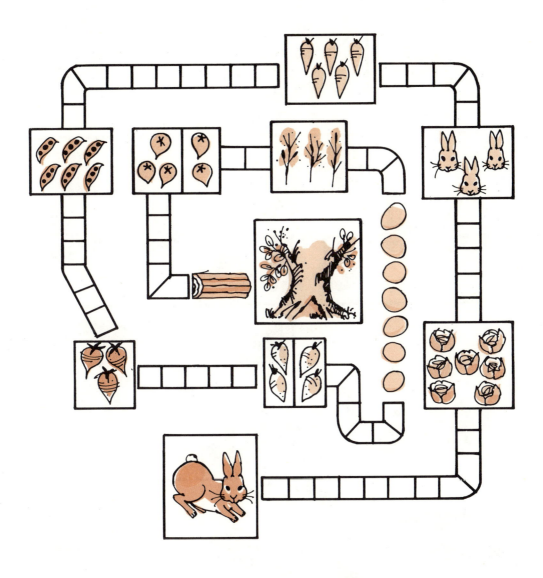

1 2 3 4 5 6 7

PROBLEM SOLVING

Tell a number story for each row.

PROBLEM SOLVING

Tell a number story for each row.

1.

2.

3.

4.

REVIEWING WHAT YOU KNOW

1. 7 subtract 2 = ☐   6 subtract 4 = ☐
3 subtract 2 = ☐   7 subtract 5 = ☐
5 subtract 3 = ☐   4 subtract 2 = ☐
7 subtract 3 = ☐   6 subtract 3 = ☐

2. 5 add 2 = ☐   1 add 6 = ☐
2 add 4 = ☐   2 add 2 = ☐
5 add 1 = ☐   4 add 3 = ☐
2 add 3 = ☐   1 add 3 = ☐

3. 7 is ☐ more than 6.   5 is ☐ less than 7.
3 is ☐ less than 4.   7 is ☐ more than 5.
5 is ☐ more than 3.   6 is ☐ more than 4.

4. How many more do you need?

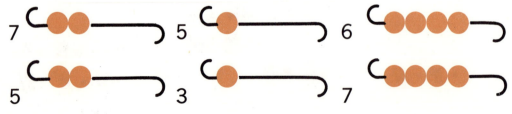

38

PRACTICE

1. How many more?

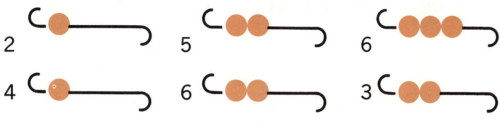

2. How many in all?

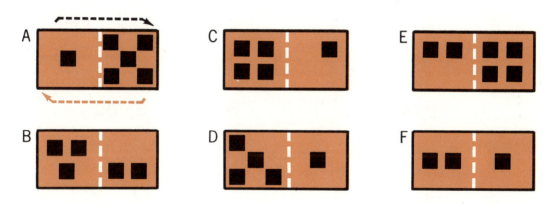

3. How many stay?

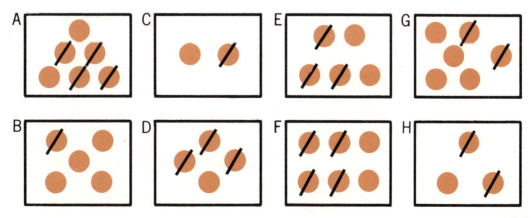

39

PRACTICE

= means **is the same as**

1. 3   add   2

3 add 2 = ☐

2   add   3

2 add 3 = ☐

2. 1   add   3

1 add 3 = ☐

3   add   1

3 add 1 = ☐

3. 4   add   1

4 add 1 = ☐

1   add   4

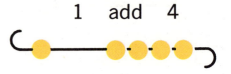

1 add 4 = ☐

4. 2   add   4

2 add 4 = ☐

4   add   2

4 add 2 = ☐

40

PRACTICE

1. 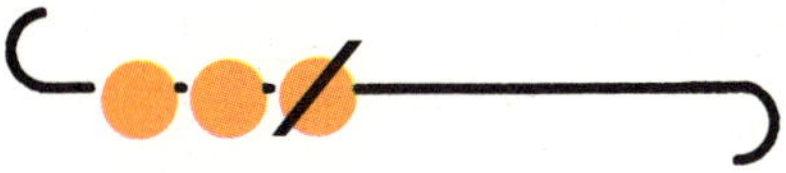

3 subtract 1 = □

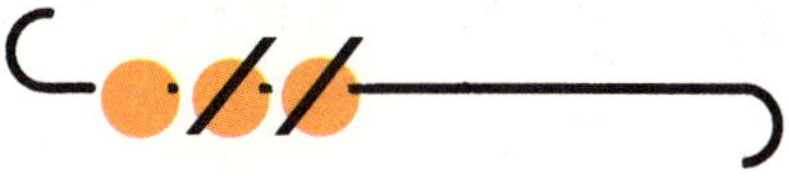

3 subtract 2 = □

2. 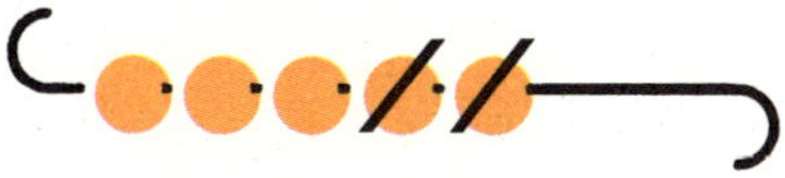

5 subtract 2 = □

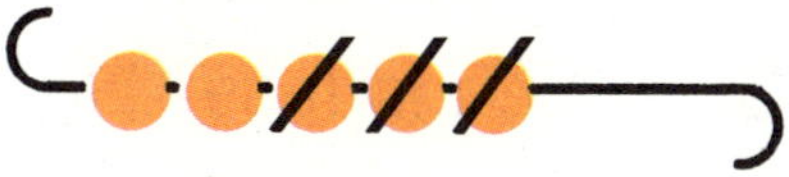

5 subtract 3 = □

3. 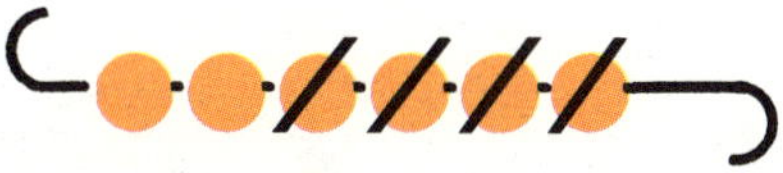

6 subtract △ = □

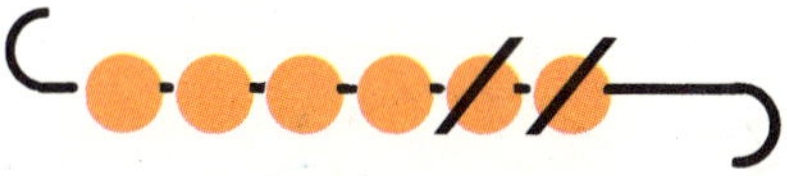

6 subtract △ = □

4. 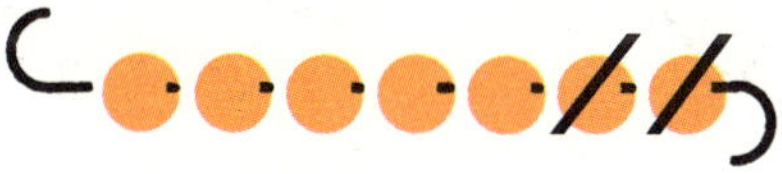

7 subtract △ = □

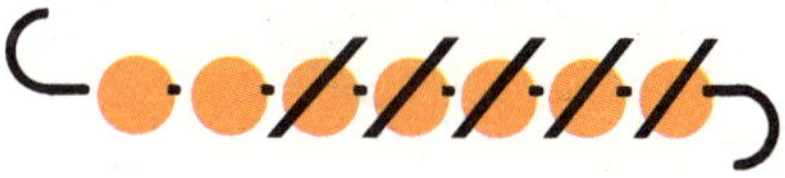

7 subtract △ = □

5. 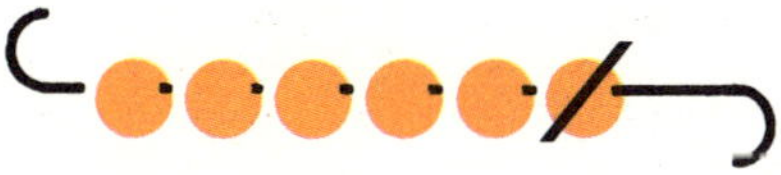

⬡ subtract △ = □

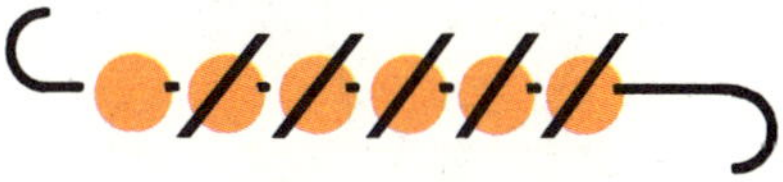

⬡ subtract △ = □

6. 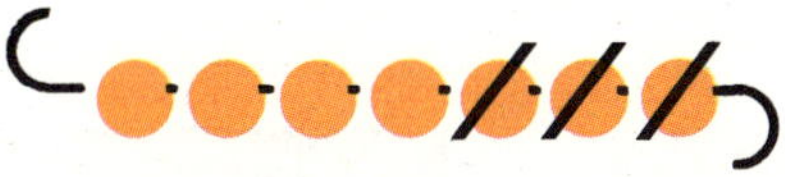

⬡ subtract △ = □

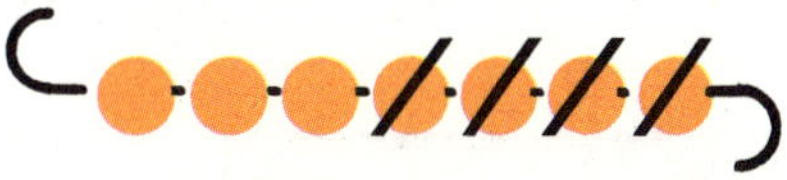

⬡ subtract △ = □

PRACTICE

1. Which is more? How many more?

A

B

C

D

E

2. Which is less? How many less?

A

B

C

D

E

3.
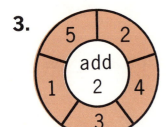
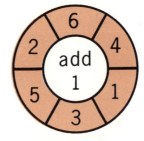
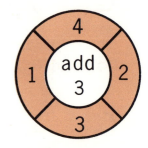

4.
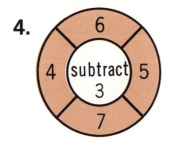
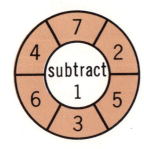
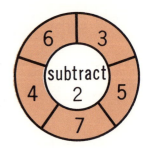

5. What numerals are missing?

A
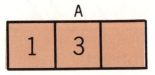

B
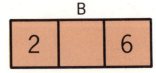

C

42

SOMETHING SPECIAL FOR YOU

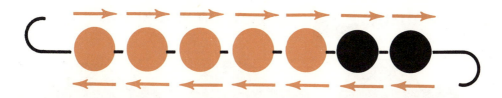

Begin on	Move	Where are you?
1	→ → → → ← ←	3
4	→ → → ← ← ← ←	☐
2	→ → → → ← ←	☐
6	→ → → → ← ← ←	☐
2	→ → → → ← ← ←	☐
5	← ← ← ← → →	☐
3	← ← → → → → →	☐
6	← ← ← ← → →	☐

HOW MANY?

1. How many 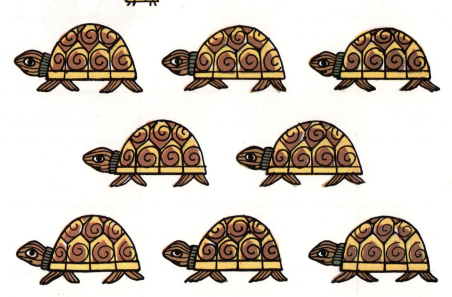's? How do you know?

2. Which set has the most eggs?

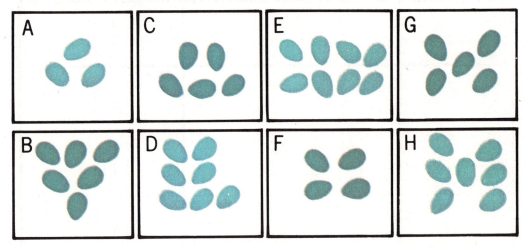

1 2 3 4 5 6 7 8

44

READING AND WRITING '8'

1. How many in each set?

2. Write a numeral for each set.

a.

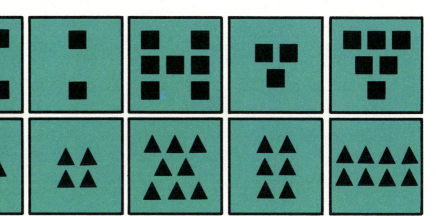

b.

3. Find a set for each numeral.

7 5 6 4 8

4. What numerals are missing?

1		3		5		7

45

HOW MANY?

1. Tell the story.

2. How many?

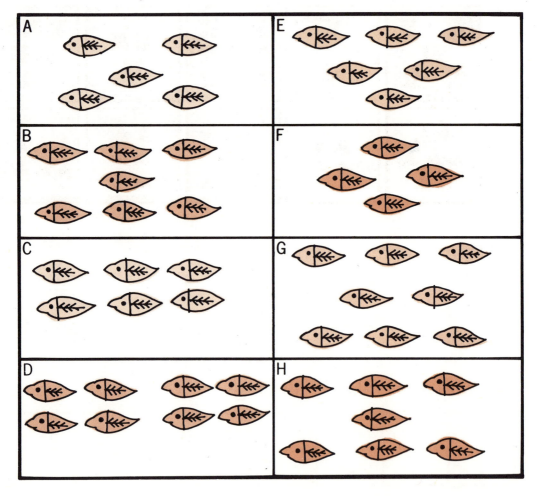

NUMBER WORDS

1. Read the number words.

2. Find two sets for each number word.

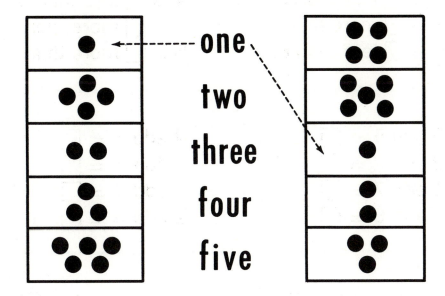

3. Read the number words in order.

four   two   five   one   three

MORE AND LESS

1. How many in each set?

A 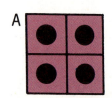 B 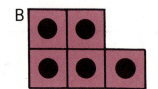 C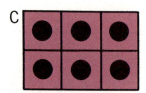

D 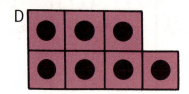 E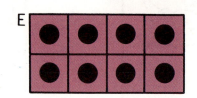

2. 4 is ☐ less than 5.   4 is ☐ less than 6.
7 is ☐ more than 6.   7 is ☐ more than 5.
6 is ☐ less than 8.   8 is ☐ more than 7.

---

3. 3 add 5 = ☐   4 add 1 = ☐
2 add 2 = ☐   2 add 3 = ☐
1 add 7 = ☐   4 add 3 = ☐
1 add 6 = ☐   2 add 5 = ☐

---

4. 7 subtract 6 = ☐   8 subtract 7 = ☐
4 subtract 3 = ☐   7 subtract 4 = ☐
8 subtract 5 = ☐   5 subtract 4 = ☐
6 subtract 5 = ☐   5 subtract 3 = ☐

HOW MANY MORE ARE NEEDED?

1. How many ●'s in each set?
How many ●'s are missing?

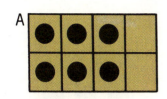 A

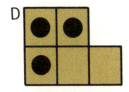 D

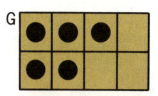 G

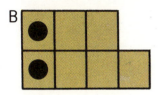 B

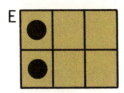 E

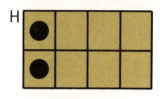 H

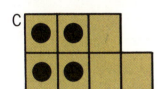 C

 F

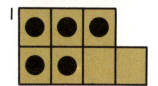 I

2.

3 add ☐ = 5   2 add ☐ = 6
5 add ☐ = 8   6 add ☐ = 8
5 add ☐ = 7   1 add ☐ = 4
2 add ☐ = 8   4 add ☐ = 6

3.

7 subtract 2 = ☐   8 subtract 6 = ☐
6 subtract 4 = ☐   7 subtract 3 = ☐
8 subtract 3 = ☐   8 subtract 2 = ☐
7 subtract 5 = ☐   6 subtract 2 = ☐

TWOS, THREES, AND FOURS

1. How many in each box?

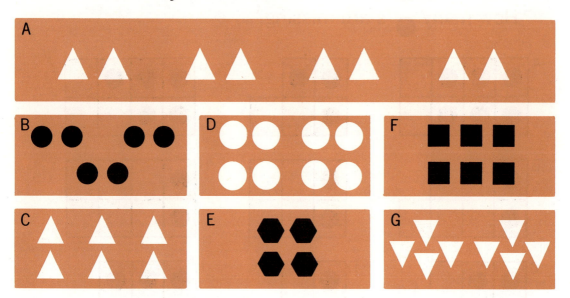

2.
2 add 2 = ☐
2 twos = ☐
4 add 4 = ☐

4 twos = ☐
3 add 3 = ☐
2 threes = ☐

3.
4 twos = ☐
2 fours = ☐
3 twos = ☐

☐ threes = 6
☐ twos = 4
☐ twos = 8

4.
☐ twos = 6
1 two = ☐

☐ fours = 8
1 four = ☐

JOINING THREE SETS

1. Tell a number story.

a. ⬤⬤/⬤ add ⬤/⬤ = △; △ add ⬤/⬤ more = □.

b. ⬤/⬤ add ⬤⬤/⬤⬤ = △; △ add ⬤/⬤ more = □.

c. ⬤ add ⬤⬤/⬤ = △; △ add ⬤⬤/⬤ more = □.

d. ⬤/⬤ add ⬤ = △; △ add ⬤⬤/⬤ more = □.

e. ⬤/⬤ add ⬤⬤/⬤ = △; △ add ⬤⬤/⬤ more = □.

f. ⬤⬤/⬤ add ⬤⬤/⬤ = △; △ add ⬤/⬤ more = □.

2.

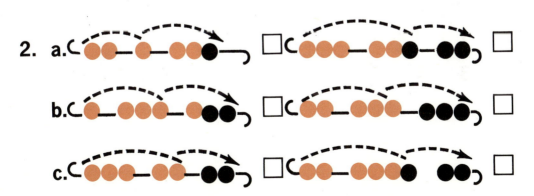

51

A SIGN FOR "ADD"

How many in each picture? Add the numbers.

☐ add ⬡

$2 + 5 = \lozenge$

$5 + 2 = \lozenge$

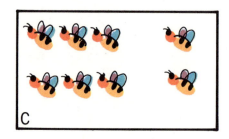

☐ add ⬡

$6 + 2 = \lozenge$

$2 + 6 = \lozenge$

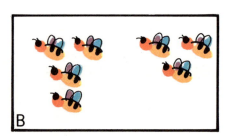

☐ add ⬡

$4 + 3 = \lozenge$

$3 + 4 = \lozenge$

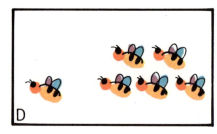

☐ add ⬡

$1 + 5 = \lozenge$

$5 + 1 = \lozenge$

A SIGN FOR "SUBTRACT"

Tell a story for each picture.

6 subtract 2 = □

6 − 2 = □   6 − 4 = ⬡

8 subtract 3 = □

8 − 3 = □   8 − 5 = ⬡

7 subtract 2 = □

7 − 2 = □   7 − 5 = ⬡

8 subtract 1 = □

8 − 1 = □   8 − 7 = ⬡

FOLLOWING THE SIGNS

2 add 3 = ⬡   8 subtract 4 = ⬡

2 + 3 = ⬡   8 − 4 = ⬡

Write the numeral.

1. 6 + 1 = ☐   8 − 3 = ☐   2 + 4 = ☐
   7 − 3 = ☐   1 + 4 = ☐   5 − 3 = ☐

2. 3 + 5 = ☐   7 − 6 = ☐   4 − 2 = ☐
   4 + 2 = ☐   8 − 2 = ☐   2 + 5 = ☐

3. 8 − 5 = ☐   5 + 2 = ☐   4 − 3 = ☐
   3 + 2 = ☐   7 − 4 = ☐   8 − 6 = ☐

4. 3 + 4 = ☐   8 − 7 = ☐   6 + 2 = ☐
   6 − 4 = ☐   4 + 4 = ☐   7 − 2 = ☐

5. 2 + 6 = ☐   7 − 5 = ☐   4 + 3 = ☐
   5 − 4 = ☐   5 + 3 = ☐   6 − 5 = ☐

THE DOUBLES

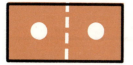  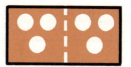 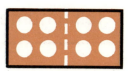

1 and 1	2 and 2	3 and 3	4 and 4
2 ones	2 twos	2 threes	2 fours

1. a. $2 + 2 = \square$ $6 - 3 = \square$ $1 + 1 = \square$

$8 - 4 = \square$ $2 - 1 = \square$ $4 + 4 = \square$

b. $4 - 2 = \square$ 2 ones $= \square$ $3 + 3 = \square$

2 fours $= \square$ 2 twos $= \square$ 2 threes $= \square$

2. Add and subtract 1.

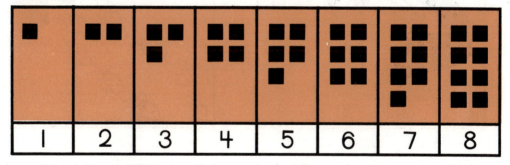

a. $4 + 1 = \square$ $1 + 4 = \square$ $4 - 1 = \square$

$6 + 1 = \square$ $1 + 6 = \square$ $6 - 1 = \square$

b. $3 + 1 = \square$ $1 + 3 = \square$ $3 - 1 = \square$

$5 + 1 = \square$ $1 + 5 = \square$ $5 - 1 = \square$

c. $7 + 1 = \square$ $1 + 7 = \square$ $7 - 1 = \square$

$2 + 1 = \square$ $1 + 2 = \square$ $2 - 1 = \square$

$1 + 1 = \square$ $8 - 1 = \square$

PROBLEM SOLVING

Tell a number story.

A

D

B

E

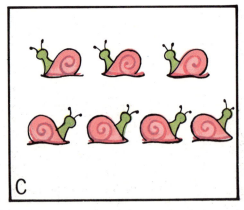

C

F

PROBLEM SOLVING

1. Tell two number stories for each picture.

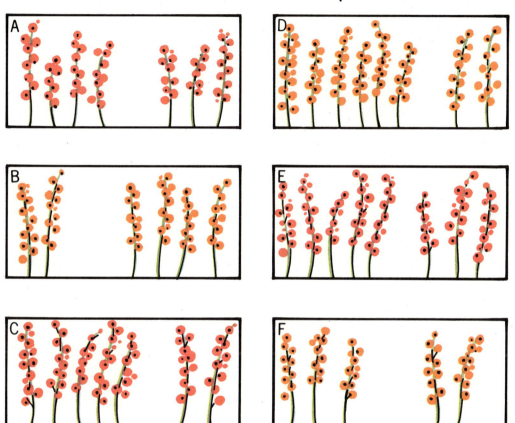

2. Find the picture for each. How many all together?

a. 6 + 2   4 + 2   2 + 5
    2 + 3   2 + 6   4 + 3

b. 3 + 2   2 + 4   3 + 4
    5 + 2   3 + 5   5 + 3

c. What other names can you find for 5, 6, 7, and 8?

PROBLEM SOLVING

Tell a number story for each row.

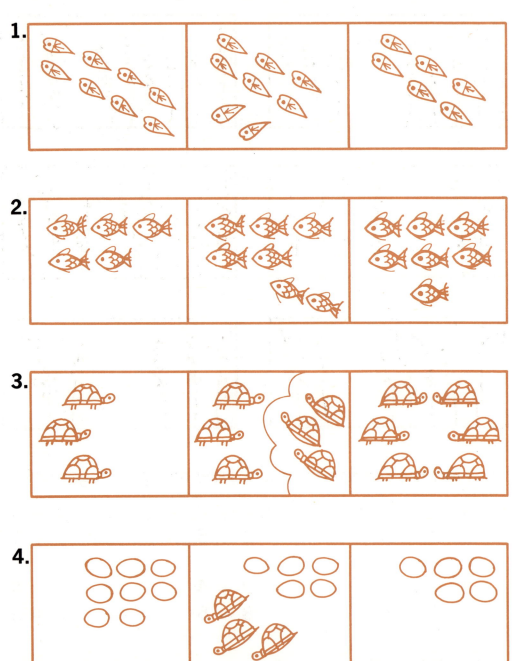

PROBLEM SOLVING

1. Tell a number story for each picture.

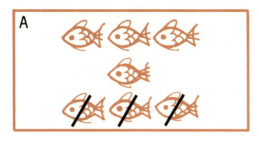

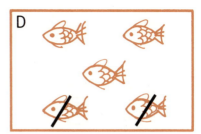

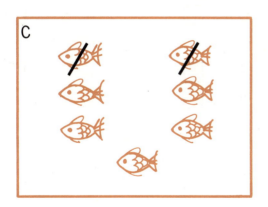

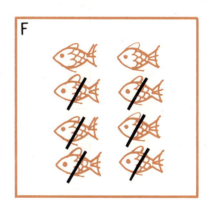

2. Find the picture for each. What is the answer?

a.	6 − 2	8 − 6	5 − 3
	7 − 3	5 − 2	8 − 5

b.	8 − 3	7 − 4	7 − 5
	7 − 2	8 − 2	6 − 4

REVIEWING WHAT YOU KNOW

1. 2 + 4 = ☐   6 + 2 = ☐   5 − 2 = ☐
5 − 4 = ☐   7 − 2 = ☐   5 + 3 = ☐

2. 3 + 2 = ☐   4 − 3 = ☐   8 − 2 = ☐
2 + 6 = ☐   8 − 3 = ☐   5 + 2 = ☐

3. 7 − 6 = ☐   5 − 3 = ☐   6 − 4 = ☐
4 + 3 = ☐   7 − 5 = ☐   2 + 3 = ☐

4. 6 − 5 = ☐   4 + 2 = ☐   7 − 4 = ☐
3 + 5 = ☐   8 − 4 = ☐   8 − 6 = ☐

5. 3 + 4 = ☐   8 − 7 = ☐   2 + 5 = ☐
6 − 2 = ☐   7 − 3 = ☐   8 − 5 = ☐

6. ☐ + 4 = 7   5 + ☐ = 8   8 − ☐ = 2
6 = ☐ + 4   7 = ☐ + 5   4 = ☐ + 3
8 = ☐ fours   6 = ☐ threes   4 = ☐ twos

7. ⠿ and ● = △; △ and ● more = ☐.

8. ● and ⠿ = △; △ and ●● more = ☐.

60

If the frames are alike, use the same numeral.

1. $7 - \square = 1$

$\square - 1 = \triangle$

$\triangle - 1 = \lozenge$

$\lozenge - 1 = 3$

2. $3 + 4 = \square$

$\square - 3 = \triangle$

$\triangle + 2 = \diamondsuit$

$\diamondsuit - 2 = 4$

3. $2 + \square = 4$

$\square + 3 = \triangle$

$\triangle + 2 = \lozenge$

$\lozenge - 2 = 5$

4. $7 - 1 = \square$

$\square - 3 = \triangle$

$\triangle + 4 = \diamondsuit$

$\diamondsuit - 4 = 3$

5. $8 - 6 = \square$ $8 - \square = 4$

$\square + 6 = 8$ $\square + \square = 8$

1. How many on A? B? C? D?

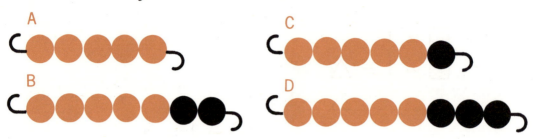

2. Which sets have the same number as A? B? C? D?

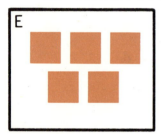

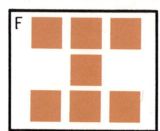

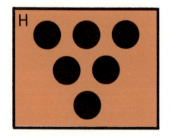

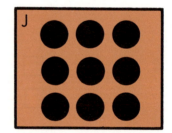

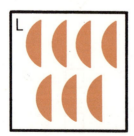

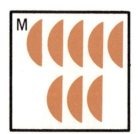

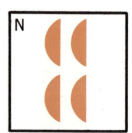

How many ▲'s in each box?

1. $3 + 4 = \triangle$ $2 + 5 = \triangle$ $4 + 4 = \triangle$
     $6 + 2 = \triangle$ $2 + 3 = \triangle$ $7 + 1 = \triangle$

2. $3 + \triangle = 8$ $1 + \triangle = 7$ $2 + \triangle = 8$
     $3 + \triangle = 6$ $2 + \triangle = 4$ $5 + \triangle = 7$

3. $6 - 4 = \triangle$ $8 - 6 = \triangle$ $7 - 2 = \triangle$
     $8 - 3 = \triangle$ $7 - 4 = \triangle$ $5 - 3 = \triangle$

4. $7 - \triangle = 4$ $6 - \triangle = 5$ $8 - \triangle = 6$
     $6 - \triangle = 4$ $4 - \triangle = 2$ $8 - \triangle = 3$

5. $7 = \triangle + 3$ $8 = 5 + \triangle$ $6 = 2 + \triangle$
     $5 = \triangle + 2$ $4 = \triangle + 1$ $8 = 1 + \triangle$

6. $2 = \triangle - 5$ $1 = \triangle - 7$ $3 = \triangle - 3$
     $4 = \triangle - 4$ $3 = \triangle - 2$ $6 = \triangle - 1$

7. 3 twos $= \triangle$ 2 ones $= \triangle$ 4 twos $= \triangle$
     2 fours $= \triangle$ 2 threes $= \triangle$ 2 twos $= \triangle$

8. $\triangle$ fours $= 8$ $\triangle$ threes $= 6$ $\triangle$ twos $= 6$
     $\triangle$ twos $= 4$ $\triangle$ twos $= 8$ $\triangle$ ones $= 2$

63

1. Write the missing numerals.

a. 4 − ⬡ = 1   6 − ⬡ = 2   6 − 3 = ⬡
6 − ⬡ = 1   3 − ⬡ = 2   6 − 4 = ⬡
5 − ⬡ = 1   7 − ⬡ = 2   6 − 2 = ⬡
7 − ⬡ = 1   5 − ⬡ = 2   6 − 1 = ⬡

b. 4 − 1 = ⬡   7 = 4 + ⬡   6 = 5 + ⬡
5 − 2 = ⬡   7 = 3 + ⬡   6 = 4 + ⬡
6 − 3 = ⬡   7 = 2 + ⬡   6 = 3 + ⬡
7 − 4 = ⬡   7 = 1 + ⬡   6 = 2 + ⬡

2. Use the same numeral for both frames.
Is the sentence true?

a. 1 + 4 = ☐   1 + 3 = ☐
       ☐ − 4 = 1   ☐ − 3 = 1

b. 5 + 2 = ☐   4 + 1 = ☐
       ☐ − 2 = 5   ☐ − 1 = 4

c. 1 + 5 = ☐   6 + 1 = ☐
       ☐ − 5 = 1   ☐ − 1 = 6

d. 2 + 2 = ☐   1 + 2 = ☐
       ☐ − 2 = 2   ☐ − 2 = 1

SOMETHING SPECIAL FOR YOU

Join the sets. How many 🔵's now?
Write the number sentence.

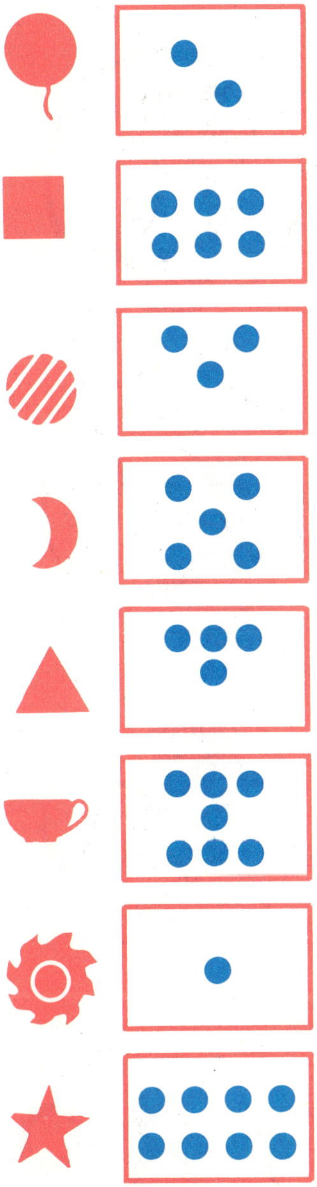

🎈 set and 🔴 set?   2 🎈 sets?

$2 + 3 = 5$ 2 twos = 4

🎈 set and 🔴 set?   2 🎈 sets?

🔴 set and ⚙ set?   2 🔴 sets?

🔴 set and ▲ set?   3 🎈 sets?

■ set and 🎈 set?   1 🎈 set?

🌙 set and 🔴 set?   4 🎈 sets?

▲ set and 🎈 set?   4 ⚙ sets?

🎈 set and 🌙 set?   1 ▲ set?

☕ set and ⚙ set?   2 ⚙ sets?

🔴 set and 🎈 set and ⚙ set?

▲ set and 🔴 set and ⚙ set?

🎈 set and ▲ set and ⚙ set?

65

HOW MANY?

How many 's in row 1? in row 2?

In rows 1 and 2? in rows 1, 2, and 3?

1 2 3 4 5 6 7 8 9

READING AND WRITING '9'

1. How many in each set?

2. Read the number words.

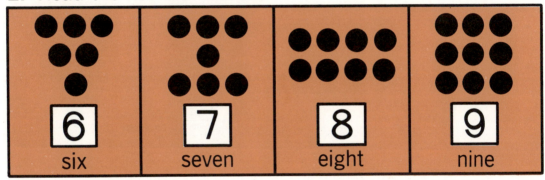

3. Match the number words, numerals, and pictures.

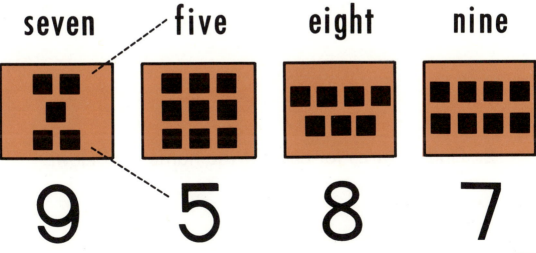

HOW MANY?

1. How many in each set in row 1? in row 2?
in row 3? in row 4? in row 5?

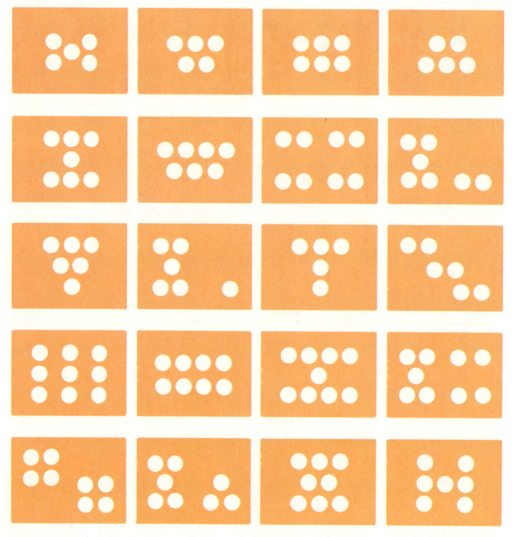

2. Find pictures for each of the numerals below.

5 7 6 8 9

WHICH ONE?

1. How many 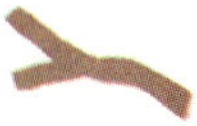 's?
2. What is on each 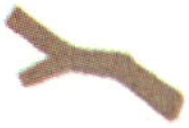 ?
3. On which 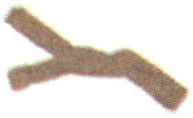 is the 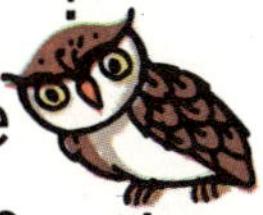 ?
   the ? the ? the ?
4. What is on 4? 6?
   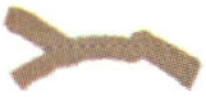 5?
5. What is on the first ?
   the third? the second?

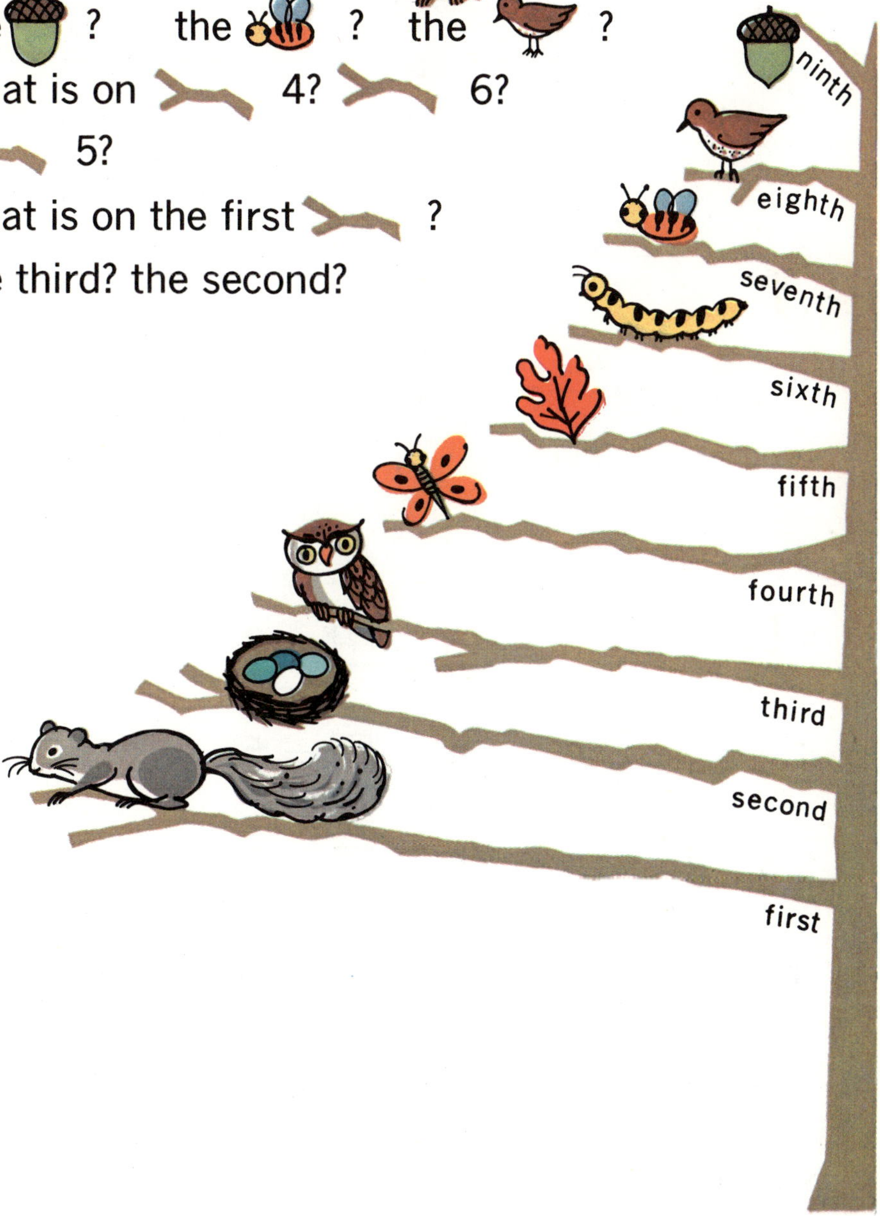

ninth
eighth
seventh
sixth
fifth
fourth
third
second
first

HOW MANY MORE?

How many in each box?
How many more are needed in each?

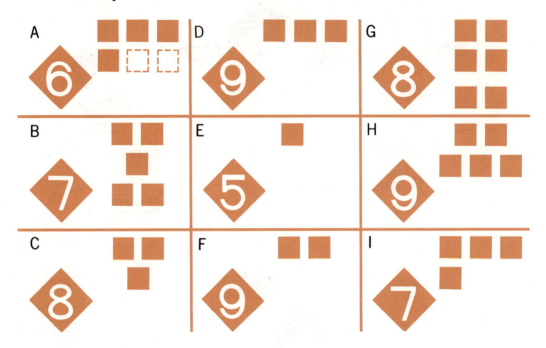

1. $3 + \square = 9$ $5 + \square = 7$ $4 + \square = 6$

2. $4 + \square = 7$ $1 + \square = 5$ $3 + \square = 8$

3. $2 + \square = 9$ $6 + \square = 8$ $5 + \square = 9$

4. $8 - 3 = \square$ $9 - 5 = \square$ $6 - 4 = \square$

5. $9 - 3 = \square$ $7 - 4 = \square$ $5 - 1 = \square$

6. $8 - 6 = \square$ $9 - 2 = \square$ $7 - 5 = \square$

HOW MANY?

Tell number stories for each bead line.

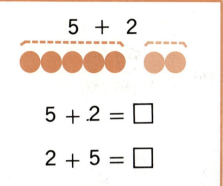

$5 + 2 = \square$

$2 + 5 = \square$

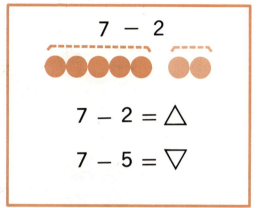

$7 - 2 = \triangle$

$7 - 5 = \triangledown$

A 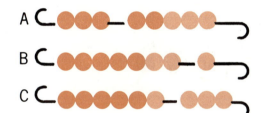   D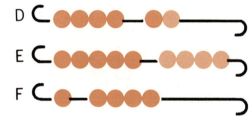

B

C

E

F

1. $4 + 2 = \square$ $6 + 3 = \square$ $1 + 4 = \square$

     $3 + 5 = \square$ $7 + 1 = \square$ $5 + 4 = \square$

2. $4 + 1 = \square$ $5 + 3 = \square$ $3 + 6 = \square$

     $4 + 5 = \square$ $2 + 4 = \square$ $1 + 7 = \square$

3. $8 - 1 = \square$ $9 - 3 = \square$ $5 - 1 = \square$

     $9 - 5 = \square$ $6 - 2 = \square$ $8 - 5 = \square$

4. $5 - 4 = \square$ $9 - 6 = \square$ $6 - 4 = \square$

     $8 - 3 = \square$ $9 - 4 = \square$ $8 - 7 = \square$

JOINING THREE SETS

1. Tell a number story for each row.

A  and and

B  and and

C  and and

D  and and

2. Add three numbers.

    a. $4 + 2 = \square$; $\square + 3 = \hexagon$.

    b. $3 + 2 = \square$; $\square + 4 = \hexagon$.

    c. $4 + 3 = \square$; $\square + 2 = \hexagon$.

    d. $2 + 3 = \square$; $\square + 2 = \hexagon$.

JOINING MORE SETS

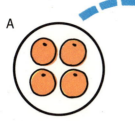  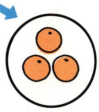

4 add 1 = $\diagup\!\!\!\square$; $\diagup\!\!\!\square$ add 3 more = $\square$.

4 + 1 = $\diagup\!\!\!\square$; $\diagup\!\!\!\square$ + 3 = $\square$.   4 + 1 + 3 = $\square$

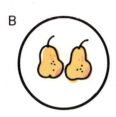  

2 add 5 = $\diagup\!\!\!\square$; $\diagup\!\!\!\square$ add 2 more = $\square$.

2 + 5 = $\diagup\!\!\!\square$; $\diagup\!\!\!\square$ + 2 = $\square$.   2 + 5 + 2 = $\square$

1. 5 + 1 + 2 = $\square$ 3 + 5 + 1 = $\square$

2. 1 + 2 + 6 = $\square$ 3 + 3 + 1 = $\square$

3. 2 + 4 + 3 = $\square$ 2 + 2 + 5 = $\square$

4. 3 + 4 + 1 = $\square$ 1 + 4 + 4 = $\square$

COUNTING BY 2's, 3's, AND 4's

1. How many in **A**? in **B**?
   in **C**? in **D**? in **E**?

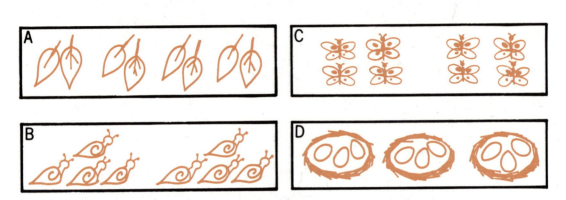

2. How many in **F**? **G**? **H**? **I**?

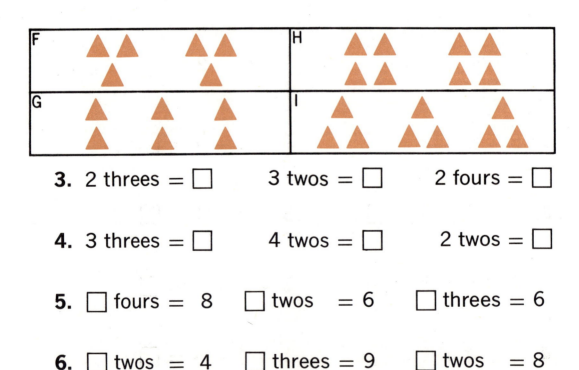

3. 2 threes = ☐   3 twos = ☐   2 fours = ☐

4. 3 threes = ☐   4 twos = ☐   2 twos = ☐

5. ☐ fours = 8   ☐ twos = 6   ☐ threes = 6

6. ☐ twos = 4   ☐ threes = 9   ☐ twos = 8

LEARNING TO MEASURE

Tell what each picture shows.

		10 20 30 40 50 60 70 80
1 2 3 4 5 6	3/4 1/2 1/4	
	PINT	36 24 12
December S M T W T F S 1 2 3 4 5 6 7 8 9 10 11 12 13 14 15 16 17 18 19 20 21 22 23 24/31 25 26 27 28 29 30	QUART	

PROBLEM SOLVING

Tell number stories about the sets.

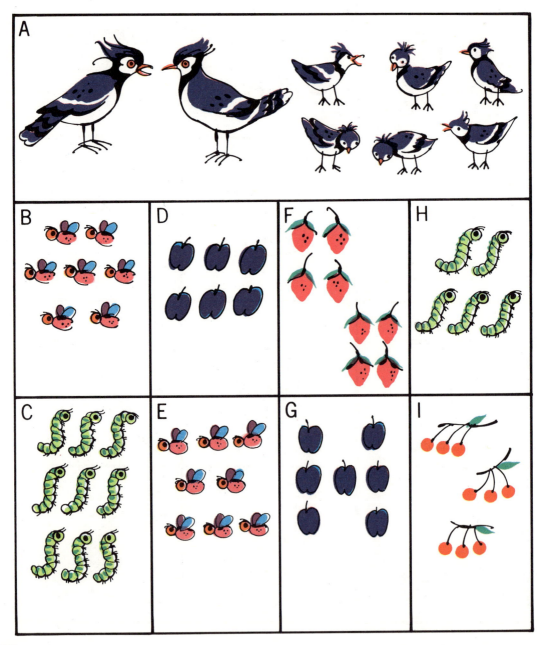

PROBLEM SOLVING

1. Tell number stories about making the 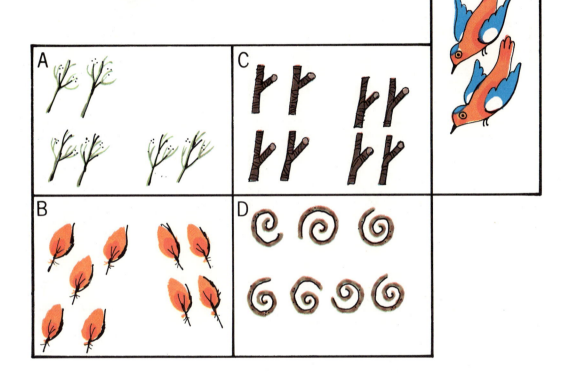 .

2. Tell number stories about these pictures.

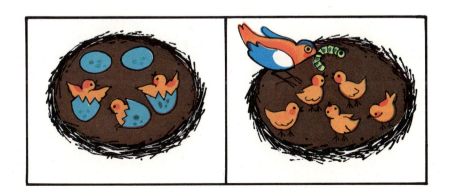

77

PROBLEM SOLVING

Tell a number story about each picture.

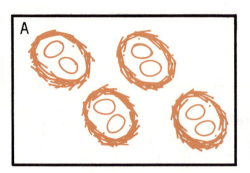

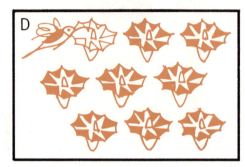

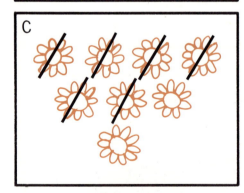

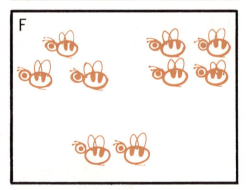

1. 7 − 4 = ☐   5 + 4 = ☐   3 twos = ☐

2. 9 − 1 = ☐   ☐ twos = 8   2 twos = ☐

3. 3 + 4 + 2 = ☐   6 + ☐ = 8

REVIEWING WHAT YOU KNOW

Use the bead line to help you.

1. $8 + 1 = \triangle$ $4 + 3 = \triangle$ $2 + 6 = \triangle$
   $3 + 5 = \triangle$ $2 + 7 = \triangle$ $1 + 4 = \triangle$
   $5 + 4 = \triangle$ $1 + 5 = \triangle$ $6 + 3 = \triangle$
   $3 + 3 = \triangle$ $5 + 2 = \triangle$ $4 + 4 = \triangle$

2. $9 = 4 + \triangle$ $5 = 2 + \triangle$ $8 = 6 + \triangle$
   $7 = 2 + \triangle$ $9 = 7 + \triangle$ $6 = 4 + \triangle$
   $9 = 3 + \triangle$ $8 = 5 + \triangle$ $7 = 3 + \triangle$

3. $6 - 4 = \triangle$ $8 - 3 = \triangle$ $9 - 6 = \triangle$
   $7 - 2 = \triangle$ $9 - 5 = \triangle$ $8 - 4 = \triangle$
   $5 - 4 = \triangle$ $7 - 4 = \triangle$ $9 - 2 = \triangle$

4. 3 twos $= \triangle$ 2 fours $= \triangle$ 3 threes $= \triangle$
   2 threes $= \triangle$ 4 twos $= \triangle$ $\triangle$ fours $= 8$
   $\triangle$ threes $= 9$ $\triangle$ twos $= 8$ $\triangle$ threes $= 6$

5. How many more or less?

 | 8 | 4 | | 9 | 7 | | 2 | 7 | | 9 | 3 |

79

CHECKING NUMBER FACTS

1. $1 + 3 = \square$ $2 + 4 = \square$ $1 + 1 = \square$
   $3 + 3 = \square$ $1 + 2 = \square$ $2 + 3 = \square$

2. $1 + 4 = \square$ $4 + 2 = \square$ $2 + 1 = \square$
   $3 + 2 = \square$ $1 + 5 = \square$ $3 + 1 = \square$

3. $5 + 1 = \square$ $2 + 2 = \square$ $4 + 1 = \square$

4. $6 - 3 = \square$ $2 - 1 = \square$ $5 - 4 = \square$
   $4 - 1 = \square$ $3 - 2 = \square$ $6 - 2 = \square$

5. $5 - 1 = \square$ $4 - 2 = \square$ $5 - 3 = \square$
   $4 - 3 = \square$ $6 - 5 = \square$ $3 - 1 = \square$

6. $6 - 1 = \square$ $5 - 2 = \square$ $6 - 4 = \square$

7. $2 + \square = 5$ $6 - \square = 1$ $1 + 4 = \square$
   $6 - 4 = \square$ $5 - 3 = \square$ $2 + \square = 6$

8. $1 + 5 = \square$ $5 - 4 = \square$ $3 + \square = 6$
   $2 + \square = 4$ $3 + \square = 5$ $4 - \square = 1$

9. Find the surprise.

3	5	1
1	3	5
5	1	3

3	1	2
1	2	3
2	3	1

2	4	3
4	3	2
3	2	4

PRACTICE

How many?

A

$8 + 1 = \square$

$1 + 8 = \square$

G

$2 + 6 = \square$

$6 + 2 = \square$

B

$3 + 4 = \square$

$4 + 3 = \square$

H

$4 + 5 = \square$

$5 + 4 = \square$

C

$3 + 3 = \square$

I

$5 + 1 = \square$

$1 + 5 = \square$

D

$3 + 6 = \square$

$6 + 3 = \square$

J

$4 + 4 = \square$

E

$6 + 1 = \square$

$1 + 6 = \square$

K

$7 + 2 = \square$

$2 + 7 = \square$

F

$3 + 5 = \square$

$5 + 3 = \square$

L

$2 + 3 = \square$

$3 + 2 = \square$

PRACTICE

How many?

A   7 − 4 = ☐
7 − 3 = ☐

G   ⑧ − 5 = △
⑧ − 3 = △

B   9 − 1 = ☐
9 − 8 = ☐

H   ☐ − 3 = △

C   8 − 6 = ☐
8 − 2 = ☐

I   ☐ − 2 = △
☐ − 7 = △

D   7 − 1 = ☐
7 − 6 = ☐

J   ☐ − 3 = △
☐ − 2 = △

E    9 − 5 = ☐
9 − 4 = ☐

K   ☐ − 1 = △
☐ − 5 = △

F   8 − 4 = ☐

L   ☐ − 6 = △
☐ − 3 = △

82

PRACTICE

1. $2 + \square = 9$ $3 + \square = 7$ $4 + \square = 9$
$5 + \square = 8$ $1 + \square = 6$ $3 + \square = 6$
$6 + \square = 8$ $6 + \square = 9$ $4 + \square = 8$

2. $6 + 3 = \square$ $5 + 2 = \square$ $2 + 6 = \square$
$2 + 4 = \square$ $3 + 5 = \square$ $7 + 2 = \square$
$5 + 4 = \square$ $4 + 3 = \square$ $2 + 3 = \square$

3. $8 - 5 = \square$ $9 - 7 = \square$ $7 - 3 = \square$
$9 - 6 = \square$ $6 - 5 = \square$ $8 - 6 = \square$
$7 - 5 = \square$ $9 - 4 = \square$ $5 - 3 = \square$

4. $5 + 1 + 3 = \square$ $4 + 4 + 1 = \square$
$2 + 4 + 2 = \square$ $3 + 2 + 3 = \square$

5.   2 threes $= \square$ 2 twos $= \square$ $\square$ twos $= 8$
3 threes $= \square$ 2 fours $= \square$ $\square$ twos $= 6$
$\square$ threes $= 6$ 4 twos $= \square$ $\square$ twos $= 4$
$\square$ threes $= 9$ 3 twos $= \square$ $\square$ fours $= 8$

PRACTICE

1. Add the numbers down, up, and across.

			6
	2	3	5
	3	2	
	5		4

4	3
2	4

2	1
3	5

2.
$3 + 5 = \square$
$2 + 6 = \square$
$1 + 7 = \square$

$7 - 2 = \square$
$6 - 2 = \square$
$5 - 2 = \square$

$8 - \square = 3$
$7 - \square = 2$
$6 - \square = 1$

3.

$\begin{array}{r} 8 \\ -3 \\ \hline \square \end{array}$
$\begin{array}{r} 6 \\ -4 \\ \hline \square \end{array}$
$\begin{array}{r} 7 \\ -6 \\ \hline \square \end{array}$
$\begin{array}{r} 5 \\ -3 \\ \hline \square \end{array}$
$\begin{array}{r} 8 \\ -6 \\ \hline \square \end{array}$
$\begin{array}{r} 4 \\ -2 \\ \hline \square \end{array}$

4.

$\begin{array}{r} 7 \\ -4 \\ \hline \square \end{array}$
$\begin{array}{r} 6 \\ -3 \\ \hline \square \end{array}$
$\begin{array}{r} 8 \\ -2 \\ \hline \square \end{array}$
$\begin{array}{r} 7 \\ -3 \\ \hline \square \end{array}$
$\begin{array}{r} 8 \\ -4 \\ \hline \square \end{array}$
$\begin{array}{r} 8 \\ -7 \\ \hline \square \end{array}$

5. $5 - 1 = \square$ $\qquad$ $6 - 1 = \square$ $\qquad$ $7 - 1 = \square$

Use the same numeral for both frames.
Are both sentences true?

1. $6 + \Box = 8$

$\qquad \Box + 6 = 8$

$2 + \Box = 7$

$\qquad \Box + 2 = 7$

2. $4 + \Box = 6$

$\qquad \Box + 4 = 6$

$3 + \Box = 8$

$\qquad \Box + 3 = 8$

3. $6 - \Box = 3$

$\qquad \Box - 2 = 1$

$8 - \Box = 4$

$\qquad \Box - 1 = 3$

4. $6 - \Box = 1$

$\qquad \Box - 3 = 2$

$7 - \Box = 2$

$\qquad \Box - 4 = 1$

5. $4 + \Box = 8$

$\qquad \Box + 4 = 8$

$8 - \Box = 5$

$\qquad \Box + 3 = 6$

6. $8 - \Box = 2$

$\qquad \Box - 2 = 4$

$7 - \Box = 3$

$\qquad \Box - 3 = 1$

7. $8 - \Box = 6$

$\qquad \Box + 2 = 4$

$8 - \Box = 3$

$\qquad \Box + 2 = 7$

1. Add 1 to each; add 2 to each.

4 2 1 3 5 7 6

2. Subtract 1 from each; subtract 2 from each.

6 3 4 7 5 8 9

3. Find the missing numbers.

$6 =$
  - $2 + \square$
  - $5 + \diagup\!\!\!\!\diagup$
  - $3 + \triangle$
  - $1 + \hexagon$

$7 =$
  - $4 + \hexagon$
  - $2 + \triangle$
  - $3 + \diagup\!\!\!\!\diagup$
  - $5 + \square$

$8 =$
  - $3 + \triangle$
  - $1 + \hexagon$
  - $5 + \square$
  - $2 + \diagup\!\!\!\!\diagup$

$3 =$
  - $6 - \square$
  - $8 - \triangle$
  - $7 - \diagup\!\!\!\!\diagup$
  - $5 - \hexagon$

$2 =$
  - $6 - \hexagon$
  - $5 - \square$
  - $7 - \triangle$
  - $8 - \diagup\!\!\!\!\diagup$

$4 =$
  - $8 - \diagup\!\!\!\!\diagup$
  - $5 - \hexagon$
  - $7 - \square$
  - $6 - \triangle$

Write the answers. Look for patterns.

1.
$1 + 2 + 3 = \bigcirc$
$2 + 2 + 2 = \bigcirc$
$2 + 3 + 1 = \bigcirc$
$1 + 2 + 5 = \bigcirc$

$3 + 4 + 2 = \bigcirc$
$3 + 2 + 4 = \bigcirc$
$2 + 4 + 3 = \bigcirc$
$2 + 5 + 1 = \bigcirc$

2.
$5 + 2 = 4 + \bigcirc$
$3 + 6 = 4 + \bigcirc$
$1 + 4 = 3 + \bigcirc$
$9 - 2 = 8 - \bigcirc$
$9 - 3 = 8 - \bigcirc$

$1 + 8 = 2 + \bigcirc$
$6 + 1 = 5 + \bigcirc$
$5 + 3 = \bigcirc + 2$
$9 - 6 = 8 - \bigcirc$
$9 - 4 = 8 - \bigcirc$

3.
$8 - \bigcirc = 4$
$6 - \bigcirc = 3$
$4 - \bigcirc = 2$

$9 - \bigcirc = 2$
$8 - \bigcirc = 2$
$7 - \bigcirc = 2$

$5 - \bigcirc = 2$
$7 - \bigcirc = 3$
$9 - \bigcirc = 4$

4.
$4 + 4 = \bigcirc$
$4 + 3 = \bigcirc$
$4 + 5 = \bigcirc$

$1 + 6 = \bigcirc$
$2 + 6 = \bigcirc$
$3 + 6 = \bigcirc$

$9 - 8 = \bigcirc$
$8 - 7 = \bigcirc$
$7 - 6 = \bigcirc$

5.
$5 - 2 = \bigcirc$
$7 - 4 = \bigcirc$
$9 - 6 = \bigcirc$

$9 - \bigcirc = 8$
$7 - \bigcirc = 6$
$5 - \bigcirc = 4$

$5 - \bigcirc = 2$
$6 - \bigcirc = 4$
$7 - \bigcirc = 6$

SOMETHING SPECIAL FOR YOU

□ □ If the frames are the same, use the same number.

△ □ If the frames are different, you may use different numbers.

1.
$$\square + \square = 8$$
$$9 = \triangle + \square$$
$$3 = \triangle - \square$$
$$\square - \triangle = 2$$

2.
$$\triangle - \square = 5$$
$$6 = \triangle + \triangle$$
$$\square + \triangle = 7$$
$$1 = \square - \triangle$$

3.
$$\square - \triangle = 8$$
$$2 = \square + \square$$
$$\square + \triangle = 4$$
$$6 = \triangle + \square$$

4. $5 + \triangle + \triangle - \square = 3$ $\square - \triangle = 6$

5. $\square + \triangle + \triangle - \square = 4$ $8 = \triangle + \square$

6. $\square + \square + \square + \square = 8$ $4 = \square - \triangle$

7. $9 - \square - \triangle - \square = 2$ $\triangle + \square = 5$

8. $\square + \square - \triangle - \triangle = 4$ $7 = \square - \triangle$

9. $8 - \square + \triangle + \triangle = 8$ $\square + \square = 4$

88

HOW MANY?

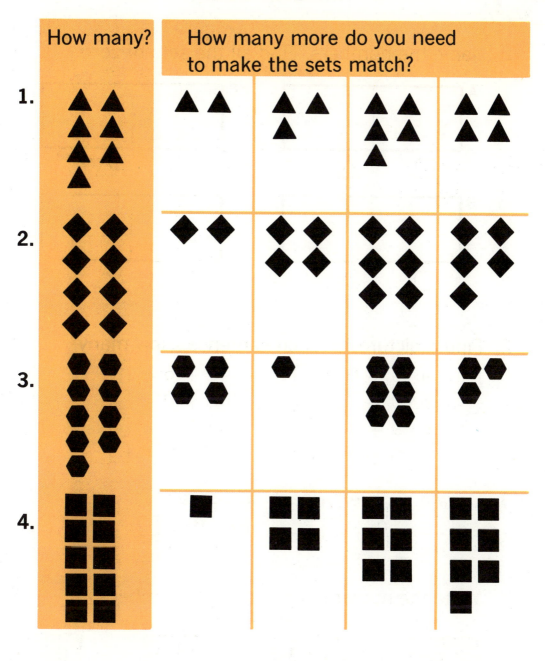

How many?	How many more do you need to make the sets match?			
1.				
2.				
3.				
4.				

1 2 3 4 5 6 7 8 9 10

ANOTHER NUMERAL

1. Name the numbers. Find all the sets of zero.

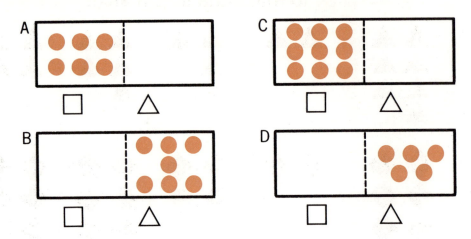

A □ △

C □ △

B □ △

D □ △

2. Find a picture for each sentence. How many?

 a. $6 + 0 = \square$ $0 + 6 = \square$
 b. $0 + 7 = \square$ $7 + 0 = \square$
 c. $9 + 0 = \square$ $0 + 9 = \square$
 d. $0 + 5 = \square$ $5 + 0 = \square$

3. Draw a picture for each sentence.

 $3 - 0 = \square$ $8 - 0 = \square$
 $5 - 0 = \square$ $4 - 0 = \square$

4. Write the numeral for the larger number.

2	3
0	5

4	1
7	5

3	0
4	8

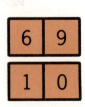

6	9
1	0

90

LEARNING ABOUT TEN

1. How many?

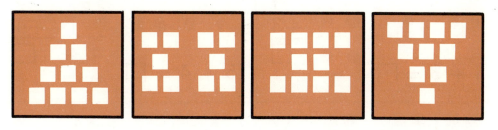

2. Read the number words. Write the numerals.

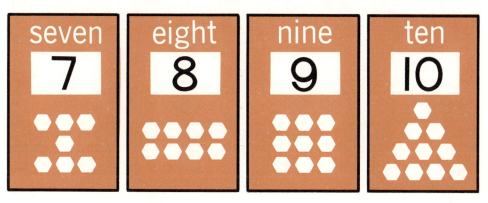

seven   7

eight   8

nine   9

ten   10

3. What do the pictures show?

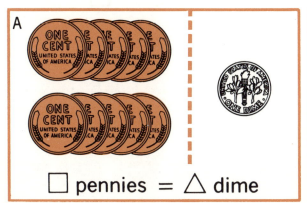

A

☐ pennies = △ dime

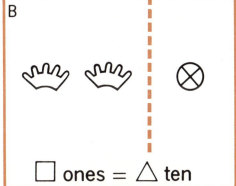

B

☐ ones = △ ten

ADDING

1.

A.
$4 + 3 = \square$

$\begin{array}{r} 4 \\ +3 \\ \hline \square \end{array}$

C.
$5 + 4 = \square$

$\begin{array}{r} 5 \\ +4 \\ \hline \square \end{array}$

B.
$6 + 2 = \square$

$\begin{array}{r} \triangledown \\ +2 \\ \hline \square \end{array}$

D.
$5 + 1 = \square$

$\begin{array}{r} 5 \\ +\triangledown \\ \hline \square \end{array}$

2.

$\begin{array}{r} 8 \\ +1 \\ \hline \square \end{array}$
$\begin{array}{r} 3 \\ +3 \\ \hline \square \end{array}$
$\begin{array}{r} 3 \\ +6 \\ \hline \square \end{array}$
$\begin{array}{r} 6 \\ +1 \\ \hline \square \end{array}$
$\begin{array}{r} 4 \\ +5 \\ \hline \square \end{array}$
$\begin{array}{r} 5 \\ +3 \\ \hline \square \end{array}$

3.

$\begin{array}{r} 3 \\ +2 \\ \hline \square \end{array}$
$\begin{array}{r} 4 \\ +4 \\ \hline \square \end{array}$
$\begin{array}{r} 2 \\ +7 \\ \hline \square \end{array}$
$\begin{array}{r} 3 \\ +5 \\ \hline \square \end{array}$
$\begin{array}{r} 1 \\ +3 \\ \hline \square \end{array}$
$\begin{array}{r} 4 \\ +2 \\ \hline \square \end{array}$

4.

$\begin{array}{r} 5 \\ +2 \\ \hline \square \end{array}$
$\begin{array}{r} 2 \\ +6 \\ \hline \square \end{array}$
$\begin{array}{r} 6 \\ +3 \\ \hline \square \end{array}$
$\begin{array}{r} 1 \\ +7 \\ \hline \square \end{array}$
$\begin{array}{r} 2 \\ +5 \\ \hline \square \end{array}$
$\begin{array}{r} 7 \\ +2 \\ \hline \square \end{array}$

92

SUBTRACTING

1.

A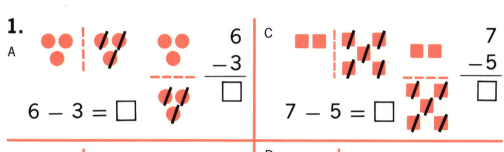

6 − 3 = ☐

$$\begin{array}{r} 6 \\ -3 \\ \hline \square \end{array}$$

C

7 − 5 = ☐

$$\begin{array}{r} 7 \\ -5 \\ \hline \square \end{array}$$

B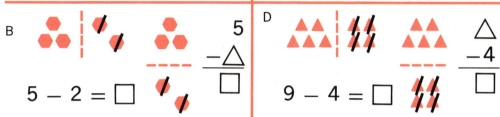

5 − 2 = ☐

$$\begin{array}{r} 5 \\ -\triangle \\ \hline \square \end{array}$$

D

9 − 4 = ☐

$$\begin{array}{r} \triangle \\ -4 \\ \hline \square \end{array}$$

2.

$$\begin{array}{r} 8 \\ -1 \\ \hline \square \end{array} \qquad \begin{array}{r} 6 \\ -2 \\ \hline \square \end{array} \qquad \begin{array}{r} 9 \\ -5 \\ \hline \square \end{array} \qquad \begin{array}{r} 7 \\ -4 \\ \hline \square \end{array} \qquad \begin{array}{r} 8 \\ -3 \\ \hline \square \end{array} \qquad \begin{array}{r} 9 \\ -2 \\ \hline \square \end{array}$$

3.

$$\begin{array}{r} 8 \\ -5 \\ \hline \square \end{array} \qquad \begin{array}{r} 9 \\ -3 \\ \hline \square \end{array} \qquad \begin{array}{r} 7 \\ -2 \\ \hline \square \end{array} \qquad \begin{array}{r} 5 \\ -4 \\ \hline \square \end{array} \qquad \begin{array}{r} 9 \\ -6 \\ \hline \square \end{array} \qquad \begin{array}{r} 6 \\ -4 \\ \hline \square \end{array}$$

4.

$$\begin{array}{r} 4 \\ -3 \\ \hline \square \end{array} \qquad \begin{array}{r} 8 \\ -2 \\ \hline \square \end{array} \qquad \begin{array}{r} 7 \\ -3 \\ \hline \square \end{array} \qquad \begin{array}{r} 9 \\ -7 \\ \hline \square \end{array} \qquad \begin{array}{r} 8 \\ -4 \\ \hline \square \end{array} \qquad \begin{array}{r} 9 \\ -1 \\ \hline \square \end{array}$$

1 2 3 4 5 6 7 8 9

LEARNING ABOUT MONEY

1¢   5¢   10¢

1. How much money?

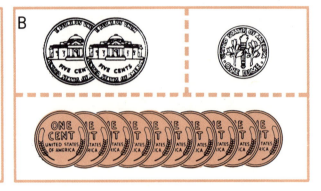

2. Which is more money? How much more?

C or    F or

D or   G  or

E  or    H  or

MORE ABOUT MONEY

How many more do you need to make the sets match?

A

B

C

D

E

1. $6¢ + 2¢ = \square ¢$ $4¢ + 3¢ = \square ¢$ $3¢ + \square ¢ = 6¢$

2. $3¢ + 6¢ = \square ¢$ $9¢ - 2¢ = \square ¢$ $4¢ + \square ¢ = 8¢$

3. $8¢ - 5¢ = \square ¢$ $10¢ - 5¢ = \square ¢$ $8¢ + \square ¢ = 10¢$

MANY WAYS TO COUNT

1. Count the beads by sets. How many beads?

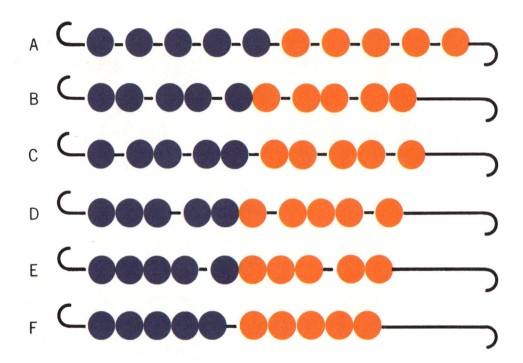

2. Write the missing numerals.

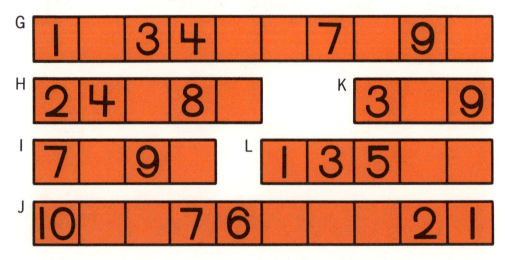

MANY WAYS TO SHOW SETS

1. Show each set in different ways.

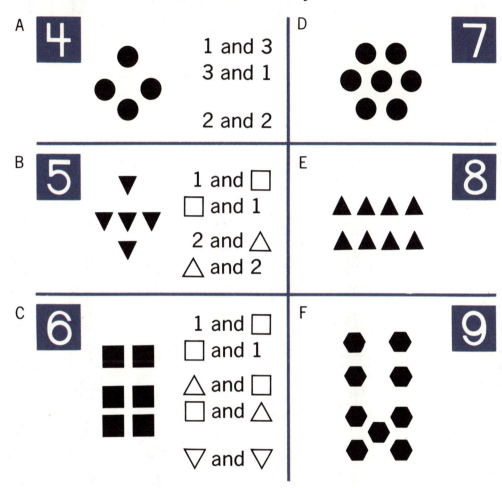

A **4**

1 and 3
3 and 1

2 and 2

D **7**

B **5**

1 and ☐
☐ and 1
2 and △
△ and 2

E **8**

C **6**

1 and ☐
☐ and 1
△ and ☐
☐ and △
▽ and ▽

F **9**

2. These sentences tell about sets of 4.

$$1 + 3 = 4 \qquad 4 - 1 = 3$$
$$3 + 1 = 4 \qquad 4 - 3 = 1$$
$$2 + 2 = 4 \qquad 4 - 2 = 2$$

3. Write number sentences for each of the other sets.

97

ADDING THREE NUMBERS

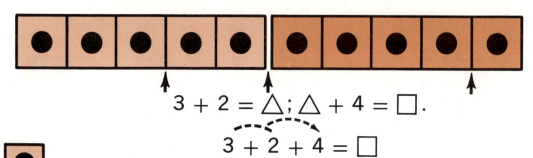

$$3 + 2 = \triangle; \triangle + 4 = \square.$$

$$3 + 2 + 4 = \square$$

1. Tell a number story for each dot line.

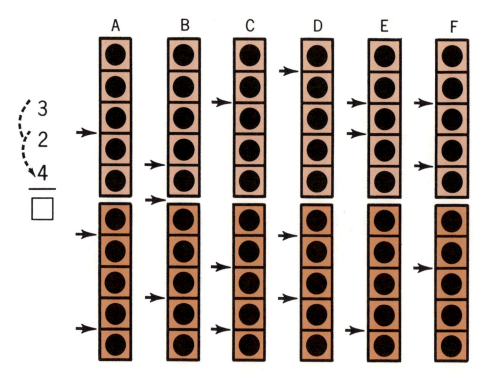

2.

4	2		3	2		2	1
1	1		3	5		2	5
3	6		3	2		3	2
$\square$	$\square$		$\square$	$\square$		$\square$	$\square$

ADDING THREE NUMBERS

1.

3	1	2	3	1
1	3	3	5	1
5	4	2	1	6
☐	☐	☐	☐	☐

2.

1	2	3	2	5
4	4	4	1	1
4	1	2	5	2
☐	☐	☐	☐	☐

3.

4	1	2	3	1
2	1	5	2	2
3	7	1	3	6
☐	☐	☐	☐	☐

4.

3	4	1	4	5
3	1	1	3	2
2	4	2	2	2
☐	☐	☐	☐	☐

5.

4	6	4	7	6
1	2	2	1	1
3	1	2	1	1
☐	☐	☐	☐	☐

USING FIVE TO ADD AND SUBTRACT

1. Write a numeral for each hand picture.

---

2. How many ●'s in each picture.

A
$$\begin{array}{c} 5 \\ +2 \\ \hline \square \end{array} \quad \begin{array}{c} \square \\ -2 \\ \hline 5 \end{array}$$

C
$$\begin{array}{c} 5 \\ +4 \\ \hline \square \end{array} \quad \begin{array}{c} \square \\ -4 \\ \hline 5 \end{array}$$

B
$$\begin{array}{c} 5 \\ +5 \\ \hline \square \end{array} \quad \begin{array}{c} \square \\ -5 \\ \hline 5 \end{array}$$

D
$$\begin{array}{c} 5 \\ +3 \\ \hline \square \end{array} \quad \begin{array}{c} \square \\ -3 \\ \hline 5 \end{array}$$

3.
$$\begin{array}{c} 5 \\ +1 \\ \hline \square \end{array} \quad \begin{array}{c} 7 \\ -2 \\ \hline \square \end{array} \quad \begin{array}{c} 4 \\ +5 \\ \hline \square \end{array} \quad \begin{array}{c} 6 \\ -1 \\ \hline \square \end{array} \quad \begin{array}{c} 2 \\ +5 \\ \hline \square \end{array} \quad \begin{array}{c} 5 \\ +3 \\ \hline \square \end{array} \quad \begin{array}{c} 9 \\ -4 \\ \hline \square \end{array}$$

4.
$$\begin{array}{c} 1 \\ +5 \\ \hline \square \end{array} \quad \begin{array}{c} 8 \\ -3 \\ \hline \square \end{array} \quad \begin{array}{c} 5 \\ +5 \\ \hline \square \end{array} \quad \begin{array}{c} 3 \\ +5 \\ \hline \square \end{array} \quad \begin{array}{c} 10 \\ -5 \\ \hline \square \end{array} \quad \begin{array}{c} 5 \\ +2 \\ \hline \square \end{array} \quad \begin{array}{c} 5 \\ +4 \\ \hline \square \end{array}$$

REVIEWING WHAT YOU KNOW

1. Which is more money? How much more?

A   or

B  or

2. How many more pennies do you need?

A

B

Find the answers.

3.

5	1	4	2	6	3	5
+4	+5	+3	+7	+2	+6	+3
☐	☐	☐	☐	☐	☐	☐

4.

7	8	9	8	6	9	9
−5	−3	−5	−6	−5	−7	−4
☐	☐	☐	☐	☐	☐	☐

5.

1	2	2	2	3	2	4
6	4	2	1	3	3	4
2	2	5	5	3	1	1
☐	☐	☐	☐	☐	☐	☐

PRACTICE

1. 3 + 1 = ☐   2 + 6 = ☐   4 + 3 = ☐
6 + 3 = ☐   2 + 2 = ☐   7 + 1 = ☐

2. 3 + 5 = ☐   6 + 1 = ☐   2 + 3 = ☐
7 + 2 = ☐   2 + 5 = ☐   1 + 4 = ☐

3. 6 + 2 = ☐   1 + 5 = ☐   5 + 4 = ☐
1 + 3 = ☐   5 + 2 = ☐   4 + 4 = ☐

4. 3 + 4 = ☐   5 + 3 = ☐   1 + 8 = ☐
4 + 2 = ☐   5 + 1 = ☐   3 + 2 = ☐

5. 3 + 6 = ☐   2 + 4 = ☐   1 + 7 = ☐
8 + 1 = ☐   3 + 3 = ☐   4 + 5 = ☐

6. 3 + 4 = ☐   2 + 7 = ☐   4 + 1 = ☐
7 + 2 = ☐   1 + 6 = ☐   6 + 3 = ☐

1. 5 − 2 = ☐   8 − 7 = ☐   6 − 3 = ☐
   4 − 1 = ☐   7 − 4 = ☐   9 − 6 = ☐

2. 8 − 2 = ☐   4 − 3 = ☐   7 − 5 = ☐
   6 − 1 = ☐   9 − 2 = ☐   8 − 6 = ☐

3. 9 − 1 = ☐   4 − 2 = ☐   5 − 3 = ☐
   7 − 3 = ☐   6 − 5 = ☐   8 − 4 = ☐

4. 5 − 4 = ☐   9 − 7 = ☐   6 − 2 = ☐
   9 − 3 = ☐   7 − 1 = ☐   8 − 5 = ☐

5. 6 − 4 = ☐   9 − 8 = ☐   7 − 2 = ☐
   8 − 1 = ☐   7 − 6 = ☐   9 − 5 = ☐

6. 5 − 1 = ☐   9 − 4 = ☐   8 − 3 = ☐
   9 − 6 = ☐   8 − 6 = ☐   7 − 5 = ☐

PRACTICE

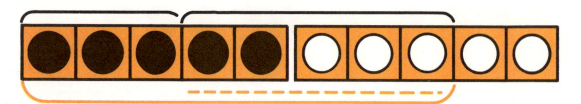

$$\begin{array}{r} 3 \\ +5 \\ \hline \square \end{array} \qquad \begin{array}{r} 8 \\ -5 \\ \hline \diagbox \end{array}$$

$3 + 5 = \square$

$8 - 5 = \square$

1.
$$\begin{array}{r} 1 \\ +8 \\ \hline \square \end{array} \quad \begin{array}{r} 2 \\ +4 \\ \hline \square \end{array} \quad \begin{array}{r} 9 \\ -8 \\ \hline \square \end{array} \quad \begin{array}{r} 6 \\ +2 \\ \hline \square \end{array} \quad \begin{array}{r} 5 \\ -3 \\ \hline \square \end{array}$$

2.
$$\begin{array}{r} 8 \\ -6 \\ \hline \square \end{array} \quad \begin{array}{r} 3 \\ +4 \\ \hline \square \end{array} \quad \begin{array}{r} 9 \\ -3 \\ \hline \square \end{array} \quad \begin{array}{r} 2 \\ +2 \\ \hline \square \end{array} \quad \begin{array}{r} 6 \\ -4 \\ \hline \square \end{array}$$

3.
$$\begin{array}{r} 2 \\ +6 \\ \hline \square \end{array} \quad \begin{array}{r} 4 \\ +5 \\ \hline \square \end{array} \quad \begin{array}{r} 9 \\ -7 \\ \hline \square \end{array} \quad \begin{array}{r} 2 \\ +3 \\ \hline \square \end{array} \quad \begin{array}{r} 3 \\ +6 \\ \hline \square \end{array} \quad \begin{array}{r} 7 \\ -3 \\ \hline \square \end{array}$$

4. $3 + 3 = \square$ $2 + 7 = \square$ $7 - 6 = \square$

5. $5 + 3 = \square$ $9 - 4 = \square$ $8 - 2 = \square$

How do the doubles help you?

$$\begin{array}{r}3\\+3\\\hline \square\end{array}$$

$$\begin{array}{r}3\\+4\\\hline \square\end{array}$$

$$\begin{array}{r}6\\-3\\\hline \square\end{array}$$

$$\begin{array}{r}6\\-2\\\hline \square\end{array}$$

$$\begin{array}{r}6\\-4\\\hline \square\end{array}$$

$$\begin{array}{r}2\\+2\\\hline \square\end{array}$$

$$\begin{array}{r}2\\+3\\\hline \square\end{array}$$

$$\begin{array}{r}4\\-2\\\hline \square\end{array}$$

$$\begin{array}{r}4\\-1\\\hline \square\end{array}$$

$$\begin{array}{r}4\\-3\\\hline \square\end{array}$$

$$\begin{array}{r}4\\+4\\\hline \square\end{array}$$

$$\begin{array}{r}4\\+5\\\hline \square\end{array}$$

$$\begin{array}{r}8\\-4\\\hline \square\end{array}$$

$$\begin{array}{r}8\\-3\\\hline \square\end{array}$$

$$\begin{array}{r}8\\-5\\\hline \square\end{array}$$

1.

$$\begin{array}{r}2\\+3\\\hline \square\end{array}\qquad\begin{array}{r}6\\-4\\\hline \square\end{array}\qquad\begin{array}{r}1\\+1\\\hline \square\end{array}\qquad\begin{array}{r}4\\-3\\\hline \square\end{array}\qquad\begin{array}{r}2\\-1\\\hline \square\end{array}\qquad\begin{array}{r}4\\+4\\\hline \square\end{array}$$

2.

$$\begin{array}{r}6\\-2\\\hline \square\end{array}\qquad\begin{array}{r}3\\+3\\\hline \square\end{array}\qquad\begin{array}{r}8\\-3\\\hline \square\end{array}\qquad\begin{array}{r}4\\+5\\\hline \square\end{array}\qquad\begin{array}{r}3\\+4\\\hline \square\end{array}\qquad\begin{array}{r}8\\-5\\\hline \square\end{array}$$

105

1. How many 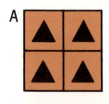's in each set?

A

B

C

D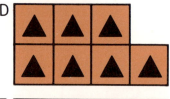

E

F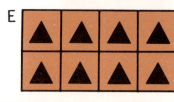

G

2.   5 is ☐ more than 4.   6 is ☐ more than 5.
      7 is ☐ more than 6.   8 is ☐ more than 7.
      9 is ☐ more than 8.   10 is ☐ more than 9.

3.   5 is ☐ less than 7.   8 is ☐ more than 6.
      9 is ☐ less than 10.   7 is ☐ less than 9.

4. 10 is ☐ more than 8.   8 is ☐ less than 9.
     7 is ☐ more than 5.   6 is ☐ less than 8.

5.   4 is ☐ less than 5.   4 is ☐ less than 6.
      9 is ☐ more than 7.   7 is ☐ less than 8.

PRACTICE

1. Find the answers.

    a. $2¢ + \square¢ = 9¢$ $\square¢ - 3¢ = 6¢$

    b. $9¢ - 5¢ = \square¢$ $8¢ - \square¢ = 1¢$

    c. $2 + \square = 8$ $\square + 3 = 9$ $3 + \square = 4$

    d. $1 = 9 - \square$ $4 = 8 - \square$ $7 = 9 - \square$

    e. $3 = 7 - \square$ $4 = 5 - \square$ $1 = 7 - \square$

2. Subtract the number in the center.

A 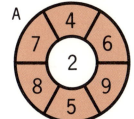   B 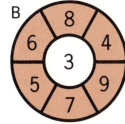   C 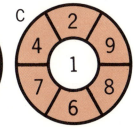   D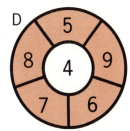

3. Add the number in the center.

A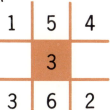

1	5	4
	3	
3	6	2

B

2+2	1+1
3	4	4
2+3	1+4

C

6	5	8
7	1	2
4	1	3

107

SOMETHING SPECIAL FOR YOU

What are the numbers?

1. Begin with 5.
Add 2.
Subtract 3.
Subtract 2.
Add 7.
The number is ☐.

4. Begin with 7.
Add 2.
Subtract 3.
Add 2.
Subtract 5.
The number is ☐.

2. Begin with 2 fours.
Subtract 3.
Add 2 twos.
Subtract 2 threes.
Add 2.
The number is ☐.

5. Begin with 4 twos.
Add 1.
Subtract 2.
Subtract 5.
Add 6.
The number is ☐.

3. Begin with 3.
Add the number
of twos in 6.
Subtract 4.
Add 7.
Subtract 5.
The number is ☐.

6. Begin with 2 threes.
Add 3.
Subtract 2 fours.
Add 8.
Subtract 3 twos.
Add 5.
The number is ☐.

1. How many?

▽ fives

☐ ten

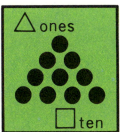
△ ones

☐ ten

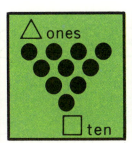
△ ones

☐ ten

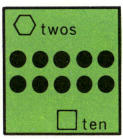
⬡ twos

☐ ten

2. How many tens in 1 row? in 2 rows? in 3 rows? Count by 10's. How many ●'s in all?

1 ten = 10
2 tens = 20
3 tens = ☐
4 tens = ☐
5 tens = ☐
6 tens = ☐
7 tens = ☐
8 tens = ☐
9 tens = ☐
10 tens = ☐

3. How many tens?

10   20   30   40   50   60   70   80   90   100

⊗   ⊗   ⊗   ⊗   ⊗   ⊗   ⊗   ⊗   ⊗   ⊗

109

MORE TENS

1. How many tens? What is the number?

A
⊗ ⊗ ⊗ ⊗ ⊗

D
⊗ ⊗ ⊗ ⊗ ⊗
⊗ ⊗

G
⊗ ⊗ ⊗

B
⊗ ⊗ ⊗ ⊗ ⊗
⊗ ⊗

E
⊗ ⊗ ⊗ ⊗ ⊗
⊗ ⊗ ⊗ ⊗

H
⊗ ⊗

I
⊗

C
⊗ ⊗ ⊗ ⊗ ⊗
⊗

F
⊗ ⊗ ⊗ ⊗ ⊗
⊗ ⊗ ⊗ ⊗ ⊗

J
⊗ ⊗ ⊗ ⊗

2.
5 tens = 50   7 tens = ☐   1 ten = ☐
2 tens = ☐   6 tens = ☐   3 tens = ☐
8 tens = ☐   4 tens = ☐   10 tens = ☐

3.
30 = 3 tens   80 = ☐ tens   50 = ☐ tens
70 = ☐ tens   100 = ☐ tens   60 = ☐ tens
90 = ☐ tens   20 = ☐ tens   40 = ☐ tens

4. Find a picture for each numeral.

10   20   30   40   50

60   70   80   90   100

COUNTING THROUGH 50

1. How many rows?

Rows	How Many ●'s
1	☐
2	☐
3	☐
4	☐
5	☐

2. Match the names.

1 row and 2 more   36

3 rows and 6 more   49

2 rows and 2 more   12

4 rows and 9 more   22

3. Write the numerals.

2 tens and 8 ones $20 + \boxed{8}$ ⬡28

1 ten and 7 ones $10 + \square$ ⬡

3 tens and 9 ones $30 + \square$ ⬡

4 tens and 0 ones $40 + \square$ ⬡

3 tens and 5 ones $30 + \square$ ⬡

2 tens and 6 ones $20 + \square$ ⬡

COUNTING TO 100

1. Count the 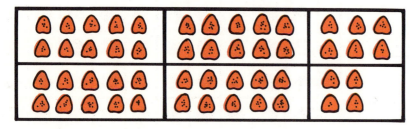 's by 2's.

    a. Joe has how many 🔺 's in all?

    b. Tom has how many in all?

    c. How many are there all together?

Joe's

Tom's

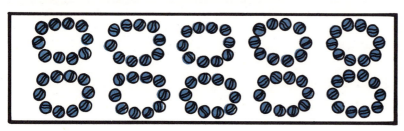

2. Count the 🔵 's by 10's.

    a. How many sets of 10 are there?

    b. How many 🔵 's?

3. Copy. Write the missing numerals.

a.	2	4	6			12			18		
b.	0	10	20		40			70			
c.	16	18			24			32			
d.		5	10	15			30			45	
e.	30	32	34		38			44			

112

10 THROUGH 20

1. How many tens? How many ones?

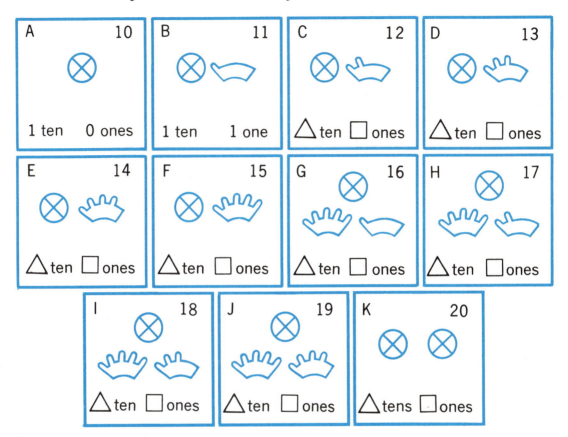

2. Match the names in each column.

10 and 1   12   10 and 6   17

10 and 2   15   10 and 7   18

10 and 3   11   10 and 8   16

10 and 4   13   10 and 9   20

10 and 5   14   10 and 10   19

TENS AND ONES

A ⊗ ✋ △ten + ◯ones = ☐	D ⊗ 👋 △ten + ◯one = ☐	G ⊗ ✋ ✋ △ten + ◯ones = ☐
B ⊗ ✋ △ten + ◯ones = ☐	E ⊗ ✋ 👋 △ten + ◯ones = ☐	H ⊗ 👋 △ten + ◯ones = ☐
C ⊗ ✋ ✋ △ten + ◯ones = ☐	F ⊗ ✋ △ten + ◯ones = ☐	I ⊗ ✋ ✋ △ten + ◯ones = ☐

1. How many tens? How many ones?
   What are the numbers?

2. a. 1 ten, 6 ones = ⬡
       1 ten, 3 ones = ⬡
   b. 15 = ☐ ten, ⬲ ones
       17 = ☐ ten, ⬲ ones

3. 10 add 1 = ⬡
   10 add 2 = ⬡
   10 add 3 = ⬡
   10 add 4 = ⬡
   10 add 5 = ⬡
   10 add 6 = ⬡
   10 add 7 = ⬡
   10 add 8 = ⬡
   10 add 9 = ⬡

4.

BEFORE		AFTER
▽14	15	△16
▽	11	△
▽	19	△
▽	13	△
▽	17	△
▽	12	△
▽	18	△

MORE TENS AND ONES

1. How many tens? How many ones?

What is the number?

A	C	E	G	I
B	D	F	H	J

2. Draw a picture for each numeral.

39   44   21   13   29

3. 1 ten, 7 ones = ⬡   3 tens, 0 ones = ⬡

3 tens, 3 ones = ⬡   4 tens, 3 ones = ⬡

2 tens, 6 ones = ⬡   3 tens, 6 ones = ⬡

4. $10 + 9 = ⬡$

$40 + 2 = ⬡$

$20 + 3 = ⬡$

$10 + 5 = ⬡$

$40 + 9 = ⬡$

$30 + 4 = ⬡$

5.

BEFORE		AFTER
28	29	▽
△	36	▽
△	19	▽
△	30	▽
△	49	▽
△	40	▽

115

ONE HALF

1. How many parts in A? B? C? D?

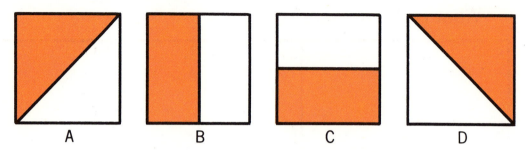

<table>
<tr><td>A</td><td>B</td><td>C</td><td>D</td></tr>
</table>

2. Which pictures have 2 parts the same size?

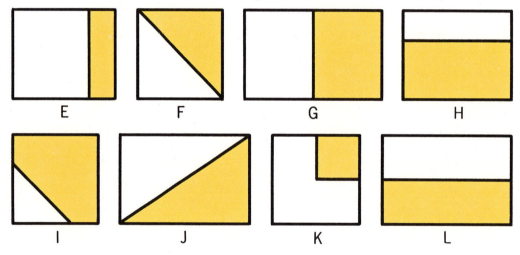

<table>
<tr><td>E</td><td>F</td><td>G</td><td>H</td></tr>
<tr><td>I</td><td>J</td><td>K</td><td>L</td></tr>
</table>

3. Show one half of each.

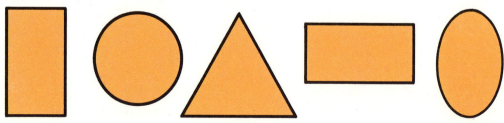

ONE FOURTH

1. How many parts in A? B? C? D?

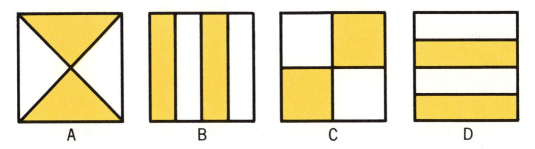

A   B   C   D

2. Which pictures have 4 parts the same size?

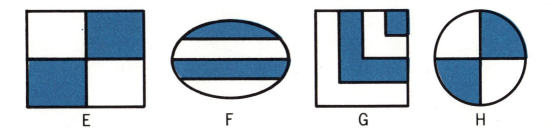

E   F   G   H

3. Which show fourths? halves?

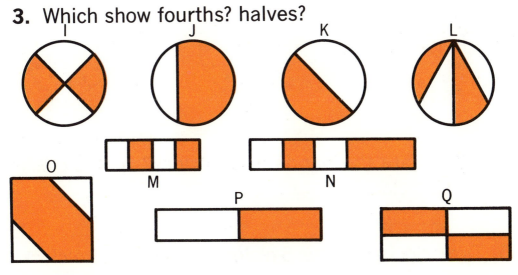

READING THE TIME

1. What time does the clock show?

2. Count by 5's from 12 to 3; from 2 to 6; from 4 to 10. Count by 5's from 12 to 12 again.

3. What time does each clock show?

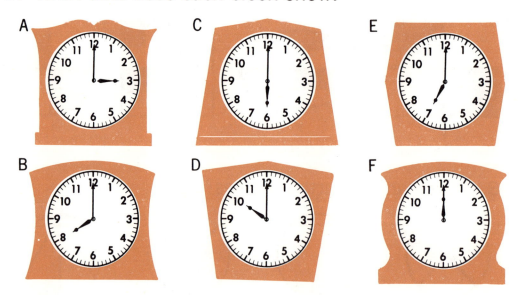

A   C   E

B   D   F

4. Where are the hands at 1 o'clock? at 4 o'clock? at 12 o'clock?

5. Draw clocks to show: 5 o'clock; 9 o'clock; 11 o'clock.

118

PROBLEM SOLVING

Tell number stories about the pictures.

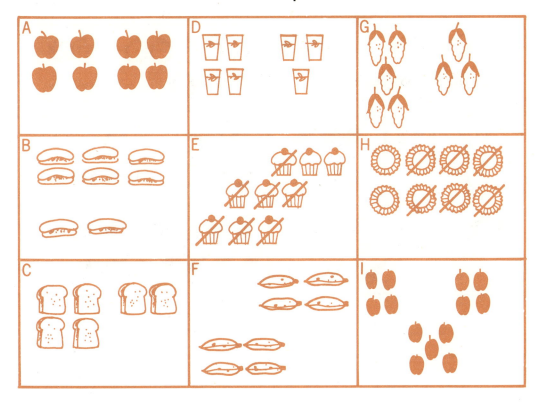

1.

4	9	2	4	8	7	9
+4	−4	+4	+3	−6	+1	−7
□	□	□	□	□	□	□

2.

6 + □ = 8   9 − 5 = □   2 fours = □

4 + □ = 9   7 − 2 = □   □ twos = 8

3 + □ = 8   8 − 3 = □   2 + 2 + 4 = □

PROBLEM SOLVING

Tell number stories about the pictures.

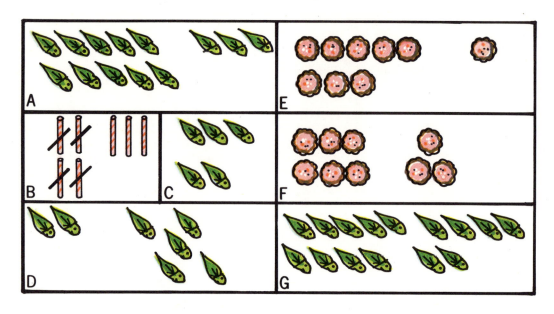

1.
$$\begin{array}{r} 3 \\ +2 \\ \hline \square \end{array} \quad \begin{array}{r} 8 \\ -5 \\ \hline \square \end{array} \quad \begin{array}{r} 6 \\ +3 \\ \hline \square \end{array} \quad \begin{array}{r} 7 \\ -4 \\ \hline \square \end{array} \quad \begin{array}{r} 6 \\ -3 \\ \hline \square \end{array} \quad \begin{array}{r} 2 \\ +5 \\ \hline \square \end{array} \quad \begin{array}{r} 9 \\ -7 \\ \hline \square \end{array}$$

2.
$3 + \square = 6 \qquad 8 - 2 = \square \qquad 7 + \square = 9$

$9 - 3 = \square \qquad 5 + \square = 8 \qquad 7 - 5 = \square$

$4 \text{ twos} = \square \qquad 2 \text{ twos} = \square \qquad 3 \text{ twos} = \square$

3.
$$\begin{array}{r} 9 \\ -\square \\ \hline 3 \end{array} \quad \begin{array}{r} 8 \\ -\square \\ \hline 2 \end{array} \quad \begin{array}{r} 6 \\ -\square \\ \hline 4 \end{array} \quad \begin{array}{r} 9 \\ -\square \\ \hline 7 \end{array} \quad \begin{array}{r} 8 \\ -\square \\ \hline 1 \end{array} \quad \begin{array}{r} 6 \\ -\square \\ \hline 2 \end{array} \quad \begin{array}{r} 9 \\ -\square \\ \hline 5 \end{array}$$

120

REVIEWING WHAT YOU KNOW

1. Write a numeral for each picture.

2. Draw pictures for these numerals.

19 50 37 12 70

3. What are the numbers?

3 tens, 3 ones = △
2 tens, 8 ones = △
6 tens, 0 ones = △
10 tens, 0 ones = △

4. Which picture shows halves? fourths?

A    B    C

5.

$$\begin{array}{ccccc} 2 & 9 & 3 & 9 & 2 \\ +3 & -2 & +4 & -5 & +6 \\ \hline \square & \square & \square & \square & \square \end{array}$$

6. What time does each clock show?

A    B    C

A

B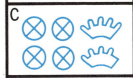

C

D

E

F

G

H

I

J

121

PRACTICE

   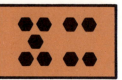

$5 + \triangle = 6$ $5 + \triangledown = 7$ $\square + \hexagon = 8$ $\square + \square\!\!\!/ = 9$

1.

$9 = \square + 4 =$
- $8 + \triangledown$
- $\hexagon + 3$
- $2 + \triangle$
- $4 + \square$

$6 = \square + 1 =$
- $2 + \triangledown$
- $\triangle + 3$
- $1 + \square$
- $4 + \hexagon$

$7 = \square + 2 =$
- $\hexagon + 1$
- $4 + \triangledown$
- $2 + \square$
- $\triangledown + 4$

$8 = \square + 3 =$
- $4 + \triangle$
- $\hexagon + 2$
- $1 + \triangledown$
- $3 + \square$

2. a.

2	9	1	7	8	7	3
$+6$	-1	$+6$	$+2$	-3	-6	$+6$
$\square$	$\square$	$\square$	$\square$	$\square$	$\square$	$\square$

b.

7	7	4	8	9	2	9
$+1$	-5	$+5$	-1	-6	$+5$	-8
$\square$	$\square$	$\square$	$\square$	$\square$	$\square$	$\square$

 1 2 3 4 5 6 7 8 9

PRACTICE

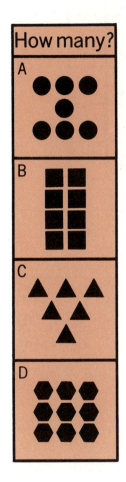

How many?	How many less?			

1.

8	7	9	6	8	9	7
−☐	−☐	−☐	−☐	−☐	−☐	−☐
4	3	6	5	5	2	5

2.

6	7	8	7	9	8	6
−☐	−☐	−☐	−☐	−☐	−☐	−☐
3	6	3	4	4	6	1

123

1	2	3	4	5	6	7	8	9	10
11	12	13	14	15	16	17	18	19	20
21	22	23	24	25	26	27	28	29	30
31	32	33	34	35	36	37	38	39	40
41	42	43	44	45	46	47	48	49	50

1. Find these numerals on the chart.

21	22	23	24	25	26	27	28	29	30
32	34	36	38	40	42	44	46	48	50
5	10	15	20	25	30	35	40	45	50

2. What are the numbers?

1 ten, 9 ones = ⑲   4 tens, 6 ones = ⬡

3 tens, 7 ones = ⬡   2 tens, 1 one = ⬡

2 tens, 9 ones = ⬡   4 tens, 4 ones = ⬡

3.

	Tens	Ones
43	△4	3
27	△	☐
38	△	☐

	Tens	Ones
48	△	☐
31	△	☐
40	△	☐

1 ten

1. How many tens?
How many ones?
What is the number?

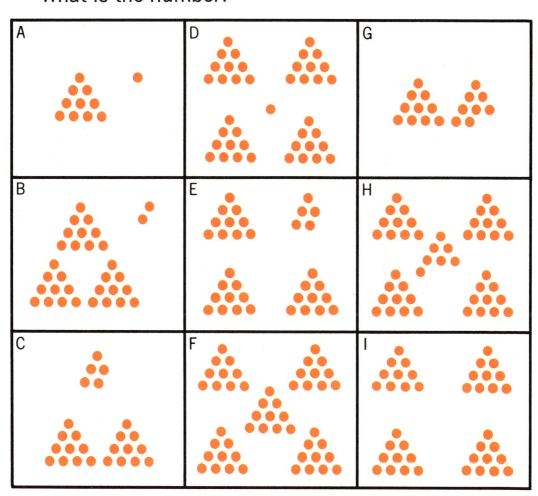

2.

$20 + 9 = \text{㉙}$ $20 + 2 = \bigcirc$

$30 + 8 = \bigcirc$ $10 + 8 = \bigcirc$

$20 + 5 = \bigcirc$ $40 + 7 = \bigcirc$

$40 + 6 = \bigcirc$ $30 + 6 = \bigcirc$

125

PRACTICE

1.

3	7	5	9	8	3	1
+6	−4	+3	−8	−5	+3	+8
□	□	□	□	□	□	□

2.
4 + 3 = □   6 − 2 = □   4 + 5 = □
9 − 3 = □   6 + 2 = □   8 − 1 = □

3.

Just Before		Just After
△	16	▽
△	47	▽
△	39	▽
△	20	▽
△	31	▽

	Tens	Ones
40	□	□
12	□	□
35	□	□
21	□	□
44	□	□

4. What are the numbers?

4 tens, 7 ones = ⬡   30 and 5 = ⬡
3 tens, 2 ones = ⬡   10 and 3 = ⬡
1 ten, 0 ones = ⬡   40 and 6 = ⬡
4 tens, 9 ones = ⬡   20 and 7 = ⬡
2 tens, 2 ones = ⬡   10 and 6 = ⬡

5. Write the numerals.

a. 16 17 ___ 19 ___ ___ 22 ___ ___ ___
b. 32 34 36 ___ ___ 42 ___ ___ ___ 50

126

1. a. Add the number at the top to each.

$4 + 5 = \square$ $4 + 3 = \square$

b. Add the number at the bottom to each.

$1 + 2 = \square$ $1 + 4 = \square$

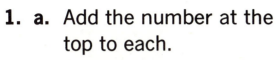

2.

7	6	5	8	5	7
+1	+2	+1	+1	+3	+2
$\square$	$\square$	$\square$	$\square$	$\square$	$\square$

3.

9	9	8	9	8	9
−7	−8	−2	−6	−3	−5
$\square$	$\square$	$\square$	$\square$	$\square$	$\square$

4.

6	4	5	5	2	3
+3	+1	+4	+2	+7	+3
$\square$	$\square$	$\square$	$\square$	$\square$	$\square$

5.

9	7	9	6	9	9
−1	−2	−4	−1	−2	−3
$\square$	$\square$	$\square$	$\square$	$\square$	$\square$

127

SOMETHING SPECIAL FOR YOU

1. Copy the numerals in the boxes in **a–e**. Write signs between the numerals to make true number sentences.

8	2	2	4

$8 - 2 = 2 + 4$

and

$8 = 2 + 2 + 4$

a. | 6 | 4 | 2 | | 2 | 3 | 5 | | 9 | 3 | 6 |

b. | 8 | 5 | 3 | | 3 | 3 | 3 | 9 | | 6 | 2 | 4 | 8 |

c. | 9 | 7 | 2 | | 7 | 3 | 2 | 2 | | 2 | 7 | 5 | 4 |

d. | 7 | 2 | 5 | | 4 | 4 | 2 | 6 | | 6 | 3 | 2 | 1 |

e. | 9 | 4 | 5 | | 2 | 3 | 1 | 6 | | 7 | 3 | 5 | 9 |

2. What number am I?

   a. I am between 20 and 40. I have 3 tens. I have as many ones as 4 add 5.

   b. My tens and ones numbers are the same. I am larger than 20 and smaller than 30.

   c. I am a teens number. I am smaller than 15. I am the same as 10 add 3.

   d. I am larger than 50 and smaller than 100. I tell you the number of minutes in an hour.

TENS AND ONES

1.

$27 = \square$ tens $+ \triangle$ ones

$35 = \square$ tens $+ \triangle$ ones

$20 = \square$ tens $+ \triangle$ ones

$13 = \square$ tens $+ \triangle$ ones

$49 = \square$ tens $+ \triangle$ ones

2. How many pennies?

4 rows and 5 more pennies $= \square$

2 rows and 9 more pennies $= \square$

1 row and 2 more pennies $= \square$

3 rows and 8 more pennies $= \square$

5 rows $= \square$ 3 rows $= \square$

MORE TENS AND ONES

1. 3 tens = ☐ ones
   5 tens = ☐ ones
   8 tens = ☐ ones
   4 tens = ☐ ones
   9 tens = ☐ ones

1 ten

10 ones

2. 30 ones = ☐ tens   30 = ☐ tens
   60 ones = ☐ tens   60 = ☐ tens
   100 ones = ☐ tens   100 = ☐ tens
   80 ones = ☐ tens   80 = ☐ tens
   50 ones = ☐ tens   50 = ☐ tens
   70 ones = ☐ tens   70 = ☐ tens

3.

Pennies	Dimes	Pennies
46 =	4 +	6
100 =	☐ +	⟋
89 =	☐ +	⟋
66 =	☐ +	⟋
21 =	☐ +	⟋
77 =	☐ +	⟋
92 =	☐ +	⟋

	Tens	Ones
46 =	4 +	6
100 =	☐ +	⟋
89 =	☐ +	⟋
66 =	☐ +	⟋
21 =	☐ +	⟋
77 =	☐ +	⟋
92 =	☐ +	⟋

HOW MANY?

1. 4 pennies + 6 pennies = ☐ pennies

6 + 4 = ☐

3 pennies + 7 pennies = ☐ pennies

7 + 3 = ☐

4 + 6 = ☐

10 − 4 = ☐

3 + 7 = ☐

10 − 3 = ☐

2.

10 − 2 = ☐	10 − 6 = ☐	10 − 5 = ☐
1 + 9 = ☐	2 + 8 = ☐	10 − 7 = ☐
2 + 4 = ☐	3 + 3 = ☐	5 + 3 = ☐

3.

4	5	5	8	6	3	9
+5	+5	+2	+2	+1	+4	+1
☐	☐	☐	☐	☐	☐	☐

4.

10	9	10	8	10	7	10
−8	−2	−3	−5	−9	−3	−1
☐	☐	☐	☐	☐	☐	☐

131

NUMBERS THROUGH ONE HUNDRED

1. How many in each row? in 10 rows?

2.
60 = ☐ tens   65 = ☐ tens + △ ones
90 = ☐ tens   89 = ☐ tens + △ ones
80 = ☐ tens   51 = ☐ tens + △ ones
50 = ☐ tens   73 = ☐ tens + △ ones

3.
9 tens + 7 ones = ☐   8 tens + 1 one = ☐
7 tens + 2 ones = ☐   9 tens + 9 ones = ☐

HOW MUCH MORE MONEY?

1. How much more money do you need?

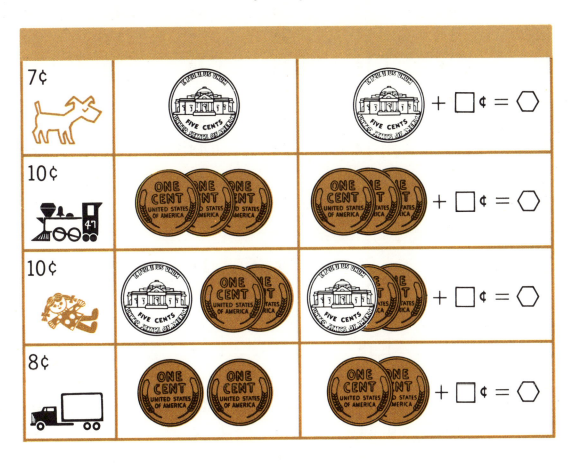

A = △
B = ▢

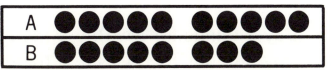

BUYING MORE THAN ONE THING

1¢	2¢	3¢	4¢	5¢

1.

1¢	3¢	2¢	1¢	3¢	4¢
4¢	2¢	2¢	4¢	4¢	2¢
1¢	1¢	2¢	2¢	1¢	3¢
☐¢	☐¢	☐¢	☐¢	☐¢	☐¢

2.

2¢	5¢	2¢	4¢	5¢	6¢
3¢	1¢	2¢	4¢	2¢	2¢
4¢	1¢	3¢	2¢	1¢	2¢
☐¢	☐¢	☐¢	☐¢	☐¢	☐¢

3.

1¢	2¢	3¢	1¢	2¢	6¢
7¢	3¢	3¢	5¢	3¢	1¢
2¢	5¢	3¢	4¢	1¢	2¢
☐¢	☐¢	☐¢	☐¢	☐¢	☐¢

4.

4¢	3¢	2¢	4¢	3¢	1¢
+5¢	+4¢	+5¢	+2¢	+5¢	+5¢
☐¢	☐¢	☐¢	☐¢	☐¢	☐¢

5. Buy 2 things for 8¢. Buy 4 things for 8¢.
Buy 3 things for 9¢. Buy 9 things for 9¢.

134

PROBLEM SOLVING

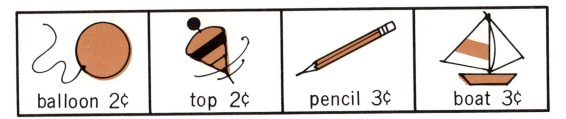

balloon 2¢   top 2¢   pencil 3¢   boat 3¢

1. What did Tom buy? What did Joe buy?

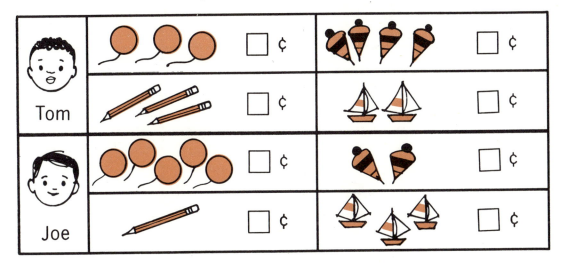

2. Make each true.

☐ fours = 8	☐ fives = 10	☐ ones = 4
☐ threes = 9	☐ twos = 8	☐ fours = 8
☐ twos = 6	☐ twos = 4	☐ threes = 9
☐ threes = 6	☐ twos = 10	☐ fives = 10

PROBLEM SOLVING

	Cost for one:	You have:	You can buy:
1.	3¢	6¢ 9¢	☐ 's ☐ 's
2.	2¢ *chocolate*	6¢ 10¢	☐ 's ☐ 's
3.	4¢	8¢ 4¢	☐ 's ☐
4.	2¢	8¢ 4¢	☐ 's ☐ 's
5.	3¢	9¢ 3¢	☐ 's ☐

6.

2 fives = ☐   10 = ☐ twos

2 threes = ☐   4 = ☐ twos

3 twos = ☐   8 = ☐ twos

2 fours = ☐   9 = ☐ threes

REVIEWING WHAT YOU KNOW

1. 3 threes = ☐   2 threes = ☐   1 five = ☐

2 fours = ☐   1 two   = ☐   5 twos = ☐

1 three = ☐   3 twos   = ☐   1 four = ☐

4 twos   = ☐   2 fives = ☐   2 twos = ☐

2. ☐ + 7 = 10   ☐ + 2 = 8   ☐ + 5 = 9

☐ + 5 = 7   ☐ + 6 = 10   ☐ + 2 = 9

3. 10 − ☐ = 2   10 − ☐ = 4   10 − 3 = ☐

8 − ☐ = 3   9 − ☐ = 7   9 − 6 = ☐

7 − ☐ = 5   8 − ☐ = 6   10 − 7 = ☐

4. 10 is ☐ more than 9.   3 is ◇ less than 7.

8 is ☐ more than 4.   5 is ⬡ less than 10.

9 is ☐ more than 3.   6 is ◇ less than 8.

5. 7¢ + 3¢ = ☐¢   10¢ − 2¢ = ☐¢

3¢ + 5¢ = ☐¢   8¢ − 4¢ = ☐¢

6. a.

$= \square$ ¢

$= \square$ ¢

b.

$= \square$ ¢

$= \square$ ¢

7. $61 = \square$ tens, $\triangle$ one $74 = \square$ tens, $\triangle$ ones

$84 = \square$ tens, $\triangle$ ones $91 = \square$ tens, $\triangle$ one

$78 = \square$ tens, $\triangle$ ones $86 = \square$ tens, $\triangle$ ones

$93 = \square$ tens, $\triangle$ ones $57 = \square$ tens, $\triangle$ ones

8.   9 tens, 7 ones $= \square$ 5 tens, 9 ones $= \square$

8 tens, 2 ones $= \square$ 1 ten,   4 ones $= \square$

10 tens, 0 ones $= \square$ 8 tens, 5 ones $= \square$

7 tens, 6 ones $= \square$ 3 tens, 9 ones $= \square$

9.

6¢	1¢	2¢	2¢	1¢	4¢	3¢
3¢	5¢	6¢	3¢	2¢	4¢	6¢
1¢	4¢	1¢	3¢	4¢	2¢	1¢
$\square$¢	$\square$¢	$\square$¢	$\square$¢	$\square$¢	$\square$¢	$\square$¢

10.

8	10	6	9	6	7	10
−7	−5	−3	−8	−6	−1	−4
$\square$	$\square$	$\square$	$\square$	$\square$	$\square$	$\square$

CHECKING NUMBER FACTS

1.
$$\begin{array}{r} 2 \\ +3 \\ \hline \square \end{array} \quad \begin{array}{r} 5 \\ +3 \\ \hline \square \end{array} \quad \begin{array}{r} 3 \\ +4 \\ \hline \square \end{array} \quad \begin{array}{r} 4 \\ +4 \\ \hline \square \end{array} \quad \begin{array}{r} 1 \\ +6 \\ \hline \square \end{array} \quad \begin{array}{r} 5 \\ +1 \\ \hline \square \end{array} \quad \begin{array}{r} 2 \\ +5 \\ \hline \square \end{array}$$

2.
$$\begin{array}{r} 1 \\ +7 \\ \hline \square \end{array} \quad \begin{array}{r} 2 \\ +6 \\ \hline \square \end{array} \quad \begin{array}{r} 4 \\ +2 \\ \hline \square \end{array} \quad \begin{array}{r} 1 \\ +4 \\ \hline \square \end{array} \quad \begin{array}{r} 4 \\ +3 \\ \hline \square \end{array} \quad \begin{array}{r} 7 \\ +1 \\ \hline \square \end{array} \quad \begin{array}{r} 2 \\ +4 \\ \hline \square \end{array}$$

3.
$$\begin{array}{r} 1 \\ +5 \\ \hline \square \end{array} \quad \begin{array}{r} 3 \\ +2 \\ \hline \square \end{array} \quad \begin{array}{r} 3 \\ +5 \\ \hline \square \end{array} \quad \begin{array}{r} 6 \\ +1 \\ \hline \square \end{array} \quad \begin{array}{r} 3 \\ +3 \\ \hline \square \end{array} \quad \begin{array}{r} 5 \\ +2 \\ \hline \square \end{array} \quad \begin{array}{r} 6 \\ +2 \\ \hline \square \end{array}$$

4.
$$\begin{array}{r} 5 \\ -4 \\ \hline \square \end{array} \quad \begin{array}{r} 7 \\ -2 \\ \hline \square \end{array} \quad \begin{array}{r} 6 \\ -4 \\ \hline \square \end{array} \quad \begin{array}{r} 8 \\ -3 \\ \hline \square \end{array} \quad \begin{array}{r} 6 \\ -2 \\ \hline \square \end{array} \quad \begin{array}{r} 7 \\ -5 \\ \hline \square \end{array} \quad \begin{array}{r} 8 \\ -7 \\ \hline \square \end{array}$$

5.
$$\begin{array}{r} 7 \\ -3 \\ \hline \square \end{array} \quad \begin{array}{r} 6 \\ -3 \\ \hline \square \end{array} \quad \begin{array}{r} 6 \\ -1 \\ \hline \square \end{array} \quad \begin{array}{r} 8 \\ -1 \\ \hline \square \end{array} \quad \begin{array}{r} 5 \\ -3 \\ \hline \square \end{array} \quad \begin{array}{r} 8 \\ -2 \\ \hline \square \end{array} \quad \begin{array}{r} 8 \\ -4 \\ \hline \square \end{array}$$

6.
$$\begin{array}{r} 7 \\ -1 \\ \hline \square \end{array} \quad \begin{array}{r} 8 \\ -5 \\ \hline \square \end{array} \quad \begin{array}{r} 5 \\ -2 \\ \hline \square \end{array} \quad \begin{array}{r} 8 \\ -6 \\ \hline \square \end{array} \quad \begin{array}{r} 6 \\ -5 \\ \hline \square \end{array} \quad \begin{array}{r} 7 \\ -4 \\ \hline \square \end{array} \quad \begin{array}{r} 7 \\ -6 \\ \hline \square \end{array}$$

PRACTICE

Let the doubles help you.

1.
$$\begin{array}{r}5\\+5\\\hline\square\end{array}\quad\begin{array}{r}5\\+4\\\hline\square\end{array}\qquad\begin{array}{r}4\\+4\\\hline\square\end{array}\quad\begin{array}{r}4\\+3\\\hline\square\end{array}\qquad\begin{array}{r}3\\+3\\\hline\square\end{array}\quad\begin{array}{r}3\\+2\\\hline\square\end{array}$$

2.
$$\begin{array}{r}2\\+2\\\hline\square\end{array}\quad\begin{array}{r}1\\+2\\\hline\square\end{array}\qquad\begin{array}{r}3\\+3\\\hline\square\end{array}\quad\begin{array}{r}2\\+3\\\hline\square\end{array}\qquad\begin{array}{r}4\\+4\\\hline\square\end{array}\quad\begin{array}{r}3\\+4\\\hline\square\end{array}$$

3.
$$\begin{array}{r}10\\-5\\\hline\square\end{array}\quad\begin{array}{r}10\\-4\\\hline\square\end{array}\qquad\begin{array}{r}8\\-4\\\hline\square\end{array}\quad\begin{array}{r}8\\-3\\\hline\square\end{array}\qquad\begin{array}{r}6\\-3\\\hline\square\end{array}\quad\begin{array}{r}6\\-2\\\hline\square\end{array}$$

4. Add 1.
$$\begin{array}{r}3\\+1\\\hline\square\end{array}\quad\begin{array}{r}4\\+1\\\hline\square\end{array}\quad\begin{array}{r}6\\+1\\\hline\square\end{array}\quad\begin{array}{r}9\\+1\\\hline\square\end{array}\quad\begin{array}{r}8\\+1\\\hline\square\end{array}\quad\begin{array}{r}5\\+1\\\hline\square\end{array}\quad\begin{array}{r}7\\+1\\\hline\square\end{array}$$

5. Subtract 1.
$$\begin{array}{r}6\\-1\\\hline\square\end{array}\quad\begin{array}{r}5\\-1\\\hline\square\end{array}\quad\begin{array}{r}8\\-1\\\hline\square\end{array}\quad\begin{array}{r}7\\-1\\\hline\square\end{array}\quad\begin{array}{r}10\\-1\\\hline\square\end{array}\quad\begin{array}{r}9\\-1\\\hline\square\end{array}\quad\begin{array}{r}4\\-1\\\hline\square\end{array}$$

PRACTICE

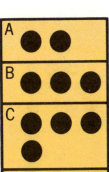

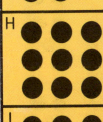

1. How many?

A = ☐   D = ☐   G = ☐

B = ☐   E = ☐   H = ☐

C = ☐   F = ☐   I = ☐

2.
$$6 + \triangle = 10 \qquad 5 + \triangle = 10$$
$$2 + \triangle = 9 \qquad 4 + \triangle = 9$$
$$2 + \triangle = 8 \qquad 7 + \triangle = 8$$
$$5 + \triangle = 9 \qquad 2 + \triangle = 7$$

3.
$$6 - \triangle = 2 \qquad 5 - \triangle = 3$$
$$9 - \triangle = 2 \qquad 10 - \triangle = 4$$
$$10 - \triangle = 7 \qquad 9 - \triangle = 6$$

4.

3	5	7	3	6	3
+5	+3	+3	+7	+3	+6
△	△	△	△	△	△

5. Find other pairs of facts.

7	2	9	1	4	2
+2	+△	+1	+△	+2	+△
▽	9	▽	10	▽	6

PRACTICE

A	B	C	D

E

Find a picture for each fact.

1.

3	4	3	2	3
+5	+3	+6	+6	+7
△	△	△	△	△

8	2	1	2	6
+2	+5	+8	+7	+2
△	△	△	△	△

F

2.

7	10	8	10	9
−6	−8	−2	−4	−4
△	△	△	△	△

G

10	9	10	8	10
−9	−3	−7	−6	−2
△	△	△	△	△

H

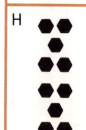

3.   9 − 6 = △   7 − 4 = △

142

PRACTICE

1.
$\square$ tens = 70
$\square$ tens = 40
$\square$ tens = 100
5 tens = $\square$
8 tens = $\square$
1 ten = $\square$
6 tens = $\square$
2 tens = $\square$
9 tens = $\square$

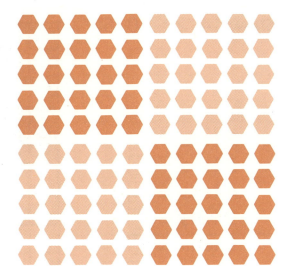

2.
6 tens + 4 ones = $\bigcirc$
5 tens + 5 ones = $\bigcirc$
1 ten + 9 ones = $\bigcirc$
9 tens + 4 ones = $\bigcirc$
8 tens + 8 ones = $\bigcirc$

83 = $\square$ tens + $\triangle$ ones
71 = $\square$ tens + $\triangle$ one
30 = $\square$ tens + $\triangle$ ones
69 = $\square$ tens + $\triangle$ ones
98 = $\square$ tens + $\triangle$ ones

3.
10¢ − 8¢ = $\square$ ¢
9¢ − 5¢ = $\square$ ¢
10¢ − 1¢ = $\square$ ¢
7¢ − 2¢ = $\square$ ¢

5¢ + 2¢ = $\square$ ¢
2¢ + 8¢ = $\square$ ¢
5¢ + 4¢ = $\square$ ¢
6¢ + 3¢ = $\square$ ¢

Show 14, 22, 70, 67.

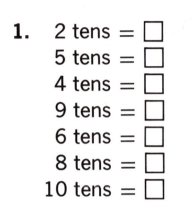

1. 2 tens = ☐
5 tens = ☐
4 tens = ☐
9 tens = ☐
6 tens = ☐
8 tens = ☐
10 tens = ☐

2. 5 tens, 3 ones = ☐   9 tens, 6 ones = ☐
9 tens, 9 ones = ☐   6 tens, 8 ones = ☐
8 tens, 7 ones = ☐   7 tens, 5 ones = ☐

3. Just before:

55?   64?   72?   90?   88?

4. Just after:

99?   56?   78?   85?   69?

5. What numerals are missing?

91   92   93   ___   ___   ___   97   ___   ___   ___

PRACTICE

1. Write the answers. Look for patterns.

$$10 - \bigcirc = 1 \qquad 2 + \bigcirc = 7 \qquad 10 - \bigcirc = 4$$
$$9 - \bigcirc = 1 \qquad 2 + \bigcirc = 8 \qquad 9 - \bigcirc = 3$$
$$8 - \bigcirc = 1 \qquad 2 + \bigcirc = 9 \qquad 8 - \bigcirc = 2$$
$$7 - \bigcirc = 1 \qquad 2 + \bigcirc = 10 \qquad 7 - \bigcirc = 1$$

2.

4	9	3	8	5	6
+6	−3	+3	−1	+2	−4
□	□	□	□	□	□

3.

9	7	9	5	8	7
−5	+3	−7	+3	−4	+2
□	□	□	□	□	□

4.

6	10	5	7	4	10
+3	−2	+1	−2	+3	−7
□	□	□	□	□	□

5.

7	5	6	10	1	9
−5	+4	−1	−4	+8	−2
□	□	□	□	□	□

6. Write the missing numerals.

$$6 + 4 = 4 + \square \qquad\qquad 10 - 4 = 9 - \square$$
$$2 + 7 = 7 + \square \qquad\qquad 9 - 3 = 8 - \square$$

1. Write the answers. Look for patterns.

$4 + 2 = \bigcirc$ $6 - 3 = \bigcirc$ $7 + \bigcirc = 8$

$6 + 2 = \bigcirc$ $8 - 3 = \bigcirc$ $8 + \bigcirc = 9$

$8 + 2 = \bigcirc$ $10 - 3 = \bigcirc$ $9 + \bigcirc = 10$

2.
$$\begin{array}{c} 6 \\ +4 \\ \hline \square \end{array} \qquad \begin{array}{c} 9 \\ -1 \\ \hline \square \end{array} \qquad \begin{array}{c} 1 \\ +6 \\ \hline \square \end{array} \qquad \begin{array}{c} 10 \\ -5 \\ \hline \square \end{array} \qquad \begin{array}{c} 4 \\ +4 \\ \hline \square \end{array} \qquad \begin{array}{c} 7 \\ -3 \\ \hline \square \end{array}$$

3.
$$\begin{array}{c} 1 \\ +9 \\ \hline \square \end{array} \qquad \begin{array}{c} 8 \\ -2 \\ \hline \square \end{array} \qquad \begin{array}{c} 3 \\ +4 \\ \hline \square \end{array} \qquad \begin{array}{c} 6 \\ -2 \\ \hline \square \end{array} \qquad \begin{array}{c} 3 \\ +6 \\ \hline \square \end{array} \qquad \begin{array}{c} 8 \\ -5 \\ \hline \square \end{array}$$

4.
$$\begin{array}{c} 7 \\ -4 \\ \hline \square \end{array} \qquad \begin{array}{c} 4 \\ +5 \\ \hline \square \end{array} \qquad \begin{array}{c} 3 \\ +7 \\ \hline \square \end{array} \qquad \begin{array}{c} 10 \\ -1 \\ \hline \square \end{array} \qquad \begin{array}{c} 1 \\ +5 \\ \hline \square \end{array} \qquad \begin{array}{c} 6 \\ +1 \\ \hline \square \end{array}$$

5.
$$\begin{array}{c} 9 \\ -4 \\ \hline \square \end{array} \qquad \begin{array}{c} 3 \\ +5 \\ \hline \square \end{array} \qquad \begin{array}{c} 2 \\ +4 \\ \hline \square \end{array} \qquad \begin{array}{c} 5 \\ +5 \\ \hline \square \end{array} \qquad \begin{array}{c} 1 \\ +7 \\ \hline \square \end{array} \qquad \begin{array}{c} 10 \\ -8 \\ \hline \square \end{array}$$

6. Write the missing numerals.

$6 - 3 = 7 - \square$ $4 + 4 = 9 - \square$

$10 - 5 = 5 - \square$ $4 + 2 = 8 - \square$

146

SOMETHING SPECIAL FOR YOU

Follow the directions.
Find the answers.

1.

	Start With	Add	Subtract	Add	Answer
a.	4	4 − 2	1	3	☐
b.	2	3 + 1	2	5	☐
c.	5	6 − 3	4	1	☐
d.	8	9 − 7	5	5	☐
e.	9	10 − 9	4	2	☐

2.

	Start With	Subtract	Subtract	Add	Answer
a.	10 − 3	3	1	2	☐
b.	6 + 3	4	2	6	☐
c.	5 + 4	2	4	4	☐
d.	7 − 4	1	1	8	☐
e.	4 + 4	3	2	1	☐

3. $5 + 3 = \square$

$\square - 2 = \hexagon$

$\hexagon + 1 = \diagup\!\!\!\square$

11, 12, and 13

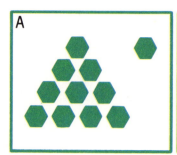 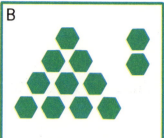 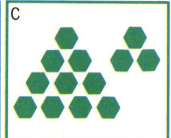

1. Picture **A**:

> $9 + 1 = 10$, so $9 + 2 = 11$.
>
> $11 - 1 = 10$, so $11 - 2 = 9$.

$10 + 1 = 11$ $10 - 1 = 9$

$2 + 9 = \square$ $11 - 9 = \square$

$8 + 3 = \square$ $7 + 4 = \square$ $6 + 5 = \square$

$11 - 5 = \square$ $11 - 4 = \square$ $11 - 3 = \square$

2. Picture **B**: $10 + 2 = \square$ $12 - 2 = \square$

$9 + 3 = \square$ $8 + 4 = \square$ $7 + 5 = \square$

$12 - 3 = \square$ $12 - 4 = \square$ $12 - 5 = \square$

3. Picture **C**: $10 + 3 = \square$ $13 - 3 = \square$

$9 + 4 = \square$ $8 + 5 = \square$ $7 + 6 = \square$

$13 - 4 = \square$ $13 - 5 = \square$ $13 - 6 = \square$

USING TEN TO FIND ANSWERS

$4 + 10 = 14,$ so $4 + 9 = 13.$

1.
$4 + 10 = \square,$ so $4 + 8 = \square.$
$3 + 10 = \square,$ so $3 + 9 = \square.$
$10 + 4 = \square,$ so $9 + 4 = \square.$
$3 + 10 = \square,$ so $3 + 8 = \square.$

2.
$12 - 10 = \square,$ so $12 - 8 = \square.$
$13 - 10 = \square,$ so $13 - 9 = \square.$
$12 - 10 = \square,$ so $12 - 9 = \square.$
$11 - 10 = \square,$ so $11 - 8 = \square.$

3.
$11 = 4 + \square$ $11 = 5 + \square$ $11 = 7 + \square$
$12 = 6 + \square$ $12 = 5 + \square$ $12 = 7 + \square$
$13 = 8 + \square$ $13 = 5 + \square$ $13 = 6 + \square$

4.
$11 - 10 = \square$ $11 - 7 = \square$ $11 - 6 = \square$
$12 - 10 = \square$ $12 - 7 = \square$ $12 - 6 = \square$
$13 - 10 = \square$ $13 - 8 = \square$ $13 - 7 = \square$

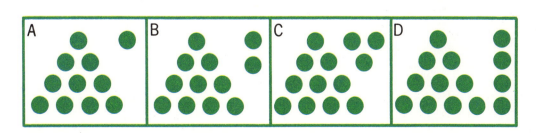

14 THROUGH 18

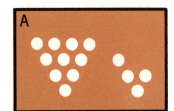

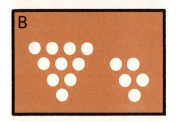

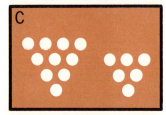

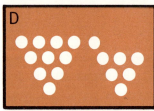

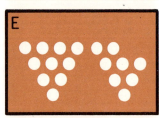

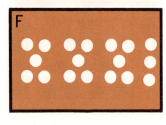

1. How many in **A**? **B**? **C**? **D**? **E**? **F**?

$10 + 5 = \square$ $15 - 5 = \square$

$10 + 4 = \square$ $14 - 4 = \square$

$10 + 6 = \square$ $16 - 6 = \square$

2. $10 + 5 = \square$ $9 + 6 = \square$

$5 + 10 = \square$ $6 + 9 = \square$

$15 - 5 = \square$ $15 - 6 = \square$

$8 + \square = 15$ $7 + \square = 15$

$15 - 7 = \square$ $15 - 8 = \square$

3. $10 + 5 = \square$ $9 + 5 = \square$

$10 + 8 = \square$ $9 + 8 = \square$

$10 + 6 = \square$ $9 + 6 = \square$

$10 + 7 = \square$ $9 + 7 = \square$

4. $18 - 8 = \square$ $18 - 9 = \square$

$14 - 4 = \square$ $14 - 5 = \square$

$16 - 6 = \square$ $16 - 7 = \square$

$17 - 7 = \square$ $17 - 8 = \square$

150

SPECIAL FACTS

1.

$$\begin{array}{r} 6 \\ +6 \\ \hline \square \end{array} \qquad \begin{array}{r} 6 \\ +\square \\ \hline 13 \end{array} \qquad \begin{array}{r} 6 \\ +5 \\ \hline \square \end{array}$$

2.

$$\begin{array}{r} 7 \\ +7 \\ \hline \square \end{array} \qquad \begin{array}{r} 7 \\ +\square \\ \hline 15 \end{array} \qquad \begin{array}{r} 7 \\ +6 \\ \hline \square \end{array}$$

3.

$$\begin{array}{r} 8 \\ +8 \\ \hline \square \end{array} \qquad \begin{array}{r} 8 \\ +\square \\ \hline 17 \end{array} \qquad \begin{array}{r} 8 \\ +7 \\ \hline \square \end{array}$$

4.

$$\begin{array}{r} 9 \\ +9 \\ \hline \square \end{array} \qquad \begin{array}{r} 9 \\ +\square \\ \hline 17 \end{array} \qquad \begin{array}{r} 8 \\ +9 \\ \hline \square \end{array}$$

5.

$$12 = \square + \square$$
$$14 = \square + \square$$
$$16 = \square + \square$$
$$18 = \square + \square$$

6.

$$\begin{array}{r} 12 \\ -6 \\ \hline \square \end{array} \qquad \begin{array}{r} 14 \\ -7 \\ \hline \square \end{array} \qquad \begin{array}{r} 16 \\ -8 \\ \hline \square \end{array} \qquad \begin{array}{r} 18 \\ -9 \\ \hline \square \end{array}$$

How many?

A

B

C

D

WORKING WITH NUMBER FACTS

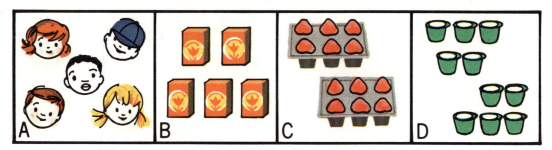

1. How many children?

How many 🥛's? How many 🟧's?

How many 🥛's for each child?

How many 🟧's for each child?

2. One 🍓 has how many 🍓's?

Two 🍓's have how many 🍓's?

Three 🍓's have how many 🍓's?

3. 2 fives = ☐ twos   3 sixes = ☐ twos

2 fours = ☐ twos   4 threes = ☐ fours

4. Count by 2's to 30.   Count by 3's to 30.

Count by 5's to 50.   Count by 10's to 50.

5. ☐ fives = 15   ☐ nines = 18

☐ threes = 12   ☐ threes = 15

☐ fours = 12   ☐ twos   = 10

☐ sixes = 18   ☐ fours = 16

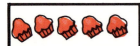   

EQUAL SETS

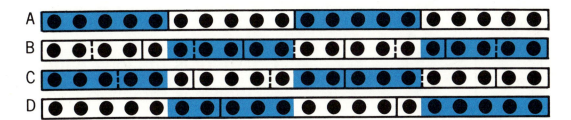

1.

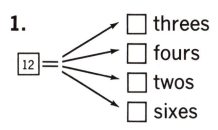

12 =
- ☐ threes
- ☐ fours
- ☐ twos
- ☐ sixes

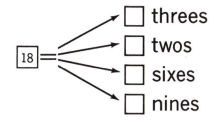

18 =
- ☐ threes
- ☐ twos
- ☐ sixes
- ☐ nines

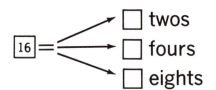

16 =
- ☐ twos
- ☐ fours
- ☐ eights

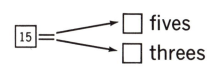

15 =
- ☐ fives
- ☐ threes

2.

10 = ⬡ twos
10 = ⬡ fives

8 = ⬡ twos
8 = ⬡ fours

6 = ⬡ threes
6 = ⬡ twos

14 = ⬡ twos
14 = ⬡ sevens

3.

2 fives = ⬡
3 fours = ⬡
6 twos = ⬡
2 eights = ⬡

6 threes = ⬡
4 fours = ⬡
2 sixes = ⬡
4 threes = ⬡

READING THE CLOCK

1. Where is the long hand?

Where is the short hand?

What time is it?

2. Which clock says 3:00?   5:00? 6:30? 10:30?

9:30? 10:00? 9:00?   4:30?

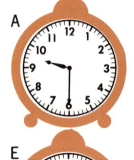

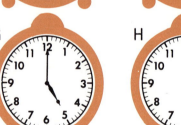

3. Where should the short hand be?

154

PROBLEM SOLVING

How many:

1. Monkeys in the tree?
   on the ground? in all?

2. Black birds? other birds? in all?

3. Four bears add 2 more = ☐ bears.
   Two bears add 4 more = ☐ bears.
   Six bears subtract 2 bears = ☐ bears.
   Six bears subtract ⬡ bears = ⬡ bears.

4. 6 monkeys and 4 bears are ☐ animals.
   3 monkeys and 3 birds and 2 bears are ☐ animals.
   2 bears and ☐ monkeys are 8 animals.

PROBLEM SOLVING

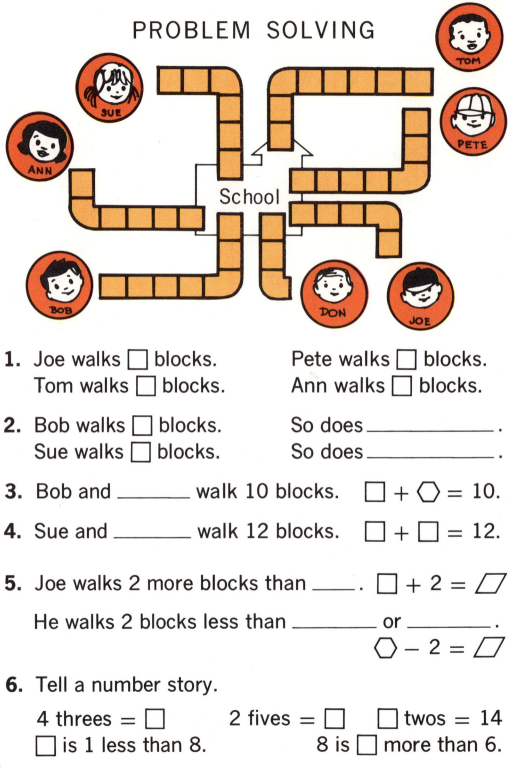

1. Joe walks ☐ blocks.   Pete walks ☐ blocks.
   Tom walks ☐ blocks.   Ann walks ☐ blocks.

2. Bob walks ☐ blocks.   So does _____ .
   Sue walks ☐ blocks.   So does _____ .

3. Bob and _____ walk 10 blocks. $\square + \hexagon = 10$.

4. Sue and _____ walk 12 blocks. $\square + \square = 12$.

5. Joe walks 2 more blocks than _____ . $\square + 2 = \diagup\!\!\!\!\diagdown$

   He walks 2 blocks less than _____ or _____ .
                                     $\hexagon - 2 = \diagup\!\!\!\!\diagdown$

6. Tell a number story.

   4 threes = ☐   2 fives = ☐   ☐ twos = 14
   ☐ is 1 less than 8.   8 is ☐ more than 6.

REVIEWING WHAT YOU KNOW

1. 8 twos = ☐   2 nines = ☐   4 fours = ☐

   5 threes = ☐   2 eights = ☐   2 sixes = ☐

   4 threes = ☐   2 sevens = ☐   3 fives = ☐

2. ☐ sixes = 18   ☐ fives = 15   ☐ nines = 18

   ☐ fours = 12   ☐ threes = 18   ☐ fives = 10

   ☐ twos = 14   ☐ sevens = 14   ☐ twos = 16

3. Which clock shows 2:30? 9:00? 11:30?

A 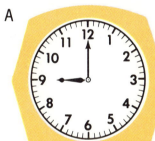   B    C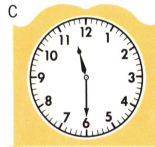

4. 10 + 2 = ☐   10 + 7 = ☐   10 + 6 = ☐

   10 + 4 = ☐   10 + 3 = ☐   10 + 9 = ☐

5. 15 − 5 = ☐   17 − 7 = ☐   16 − 6 = ☐

   14 − 10 = ☐   12 − 10 = ☐   13 − 10 = ☐

6. 7 + 4 = ☐   8 + 6 = ☐   9 + 7 = ☐

   3 + 9 = ☐   9 + 8 = ☐   8 + 7 = ☐

157

CHECKING NUMBER FACTS

1.
3	5	4	5	2	3	1
+4	+1	+6	+3	+7	+3	+9
□	□	□	□	□	□	□

2.
4	2	6	3	5	1	2
+5	+4	+1	+7	+2	+8	+6
□	□	□	□	□	□	□

3.
7	6	3	1	2	4	5
+1	+4	+5	+5	+8	+2	+4
□	□	□	□	□	□	□

4.
9	7	8	10	8	7	10
−2	−4	−1	−6	−5	−2	−9
□	□	□	□	□	□	□

5.
8	6	10	9	8	10	6
−7	−3	−4	−5	−2	−8	−1
□	□	□	□	□	□	□

6.
2	9	□	8	4	10	9
+□	−9	+3	−□	+□	−10	−□
10	□	9	5	7	□	4

CHECKING NUMBER FACTS

1. $\begin{array}{r} 8 \\ -4 \\ \hline \square \end{array}$ $\begin{array}{r} 10 \\ -7 \\ \hline \square \end{array}$ $\begin{array}{r} 6 \\ -5 \\ \hline \square \end{array}$ $\begin{array}{r} 9 \\ -4 \\ \hline \square \end{array}$ $\begin{array}{r} 6 \\ -2 \\ \hline \square \end{array}$ $\begin{array}{r} 9 \\ -1 \\ \hline \square \end{array}$ $\begin{array}{r} 7 \\ -5 \\ \hline \square \end{array}$

2. $\begin{array}{r} 10 \\ -2 \\ \hline \square \end{array}$ $\begin{array}{r} 8 \\ -6 \\ \hline \square \end{array}$ $\begin{array}{r} 9 \\ -3 \\ \hline \square \end{array}$ $\begin{array}{r} 7 \\ -1 \\ \hline \square \end{array}$ $\begin{array}{r} 6 \\ -4 \\ \hline \square \end{array}$ $\begin{array}{r} 10 \\ -3 \\ \hline \square \end{array}$ $\begin{array}{r} 9 \\ -8 \\ \hline \square \end{array}$

3. $\begin{array}{r} 8 \\ -3 \\ \hline \square \end{array}$ $\begin{array}{r} 10 \\ -1 \\ \hline \square \end{array}$ $\begin{array}{r} 7 \\ -3 \\ \hline \square \end{array}$ $\begin{array}{r} 9 \\ -7 \\ \hline \square \end{array}$ $\begin{array}{r} 7 \\ -6 \\ \hline \square \end{array}$ $\begin{array}{r} 10 \\ -5 \\ \hline \square \end{array}$ $\begin{array}{r} 9 \\ -6 \\ \hline \square \end{array}$

---

4. $\begin{array}{r} 6 \\ +2 \\ \hline \square \end{array}$ $\begin{array}{r} 9 \\ +1 \\ \hline \square \end{array}$ $\begin{array}{r} 4 \\ +3 \\ \hline \square \end{array}$ $\begin{array}{r} 3 \\ +6 \\ \hline \square \end{array}$ $\begin{array}{r} 5 \\ +5 \\ \hline \square \end{array}$ $\begin{array}{r} 1 \\ +6 \\ \hline \square \end{array}$ $\begin{array}{r} 7 \\ +2 \\ \hline \square \end{array}$

.5. $\begin{array}{r} 2 \\ +5 \\ \hline \square \end{array}$ $\begin{array}{r} 8 \\ +1 \\ \hline \square \end{array}$ $\begin{array}{r} 7 \\ +3 \\ \hline \square \end{array}$ $\begin{array}{r} 4 \\ +4 \\ \hline \square \end{array}$ $\begin{array}{r} 8 \\ +2 \\ \hline \square \end{array}$ $\begin{array}{r} 6 \\ +3 \\ \hline \square \end{array}$ $\begin{array}{r} 1 \\ +7 \\ \hline \square \end{array}$

6. $\begin{array}{r} 6 \\ +\square \\ \hline 8 \end{array}$ $\begin{array}{r} 10 \\ -\square \\ \hline 3 \end{array}$ $\begin{array}{r} 7 \\ -\square \\ \hline 2 \end{array}$ $\begin{array}{r} 4 \\ +\square \\ \hline 10 \end{array}$ $\begin{array}{r} 6 \\ -6 \\ \hline \square \end{array}$ $\begin{array}{r} \square \\ +7 \\ \hline 9 \end{array}$ $\begin{array}{r} 8 \\ -8 \\ \hline \square \end{array}$

1. $10 + 6 = \square + 10$ $10 + 8 = \square + 10$

    $5 + 10 = \square$ $5 + 9 = \square$

    $7 + 10 = \square$ $7 + 9 = \square$

2. a. $16 - 10 = \square$ $16 - 9 = \square$

   b. $15 - 10 = \square$ $15 - 9 = \square$

   c. $14 - 10 = \square$ $14 - 9 = \square$

   d. $17 - 10 = \square$ $17 - 9 = \square$

   e. $18 - 10 = \square$ $18 - 9 = \square$

3. $14 - 8 = \square$ $15 - 7 = \square$ $14 - 6 = \square$

    $15 - 8 = \square$ $16 - 8 = \square$ $14 - 7 = \square$

4. $7 + 8 = \square$ $8 + 9 = \square$ $8 + 6 = \square$

    $6 + 8 = \square$ $8 + 7 = \square$ $6 + 9 = \square$

5. $14 = 7 + \triangledown$ $16 = 8 + \triangledown$ $18 = 9 + \triangledown$

1. Count by 2's.   How many in A? B? C?

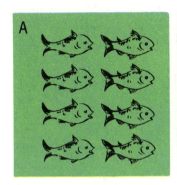 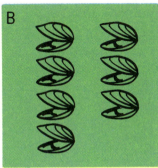 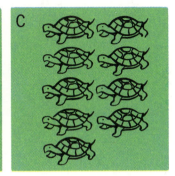

2. Look at picture A:

$$8 - 2 = \triangle \qquad \triangle - 2 = \hexagon \qquad \hexagon - 2 = \diamond$$

3.
6	3	4	7	5	2
+2	+2	+2	+2	+2	+2
□	□	□	□	□	□

4.
8	4	7	6	9	5
−2	−2	−2	−2	−2	−2
□	□	□	□	□	□

5. What numerals are missing?

2	4	___	8	___	___	___
16	___	12	10	___	___	___
3	5	7	___	11	___	15
13	11	___	___	5	___	___

161

PRACTICE

1.

9	8	5	9	6	7	6
+6	+4	+8	+5	+5	+9	+8
□	□	□	□	□	□	□

2.

11	15	13	18	12	14	11
−5	−8	−9	−9	−5	−9	−3
□	□	□	□	□	□	□

3.

14	3	11	9	11	16	4
−6	+9	−4	+2	−6	−7	+9
□	□	□	□	□	□	□

4.

$7 + 3 = \square$ $11 - 7 = \square$ $7 + \square = 14$

$8 + 3 = \square$ $12 - 7 = \square$ $8 + \square = 16$

$9 + 3 = \square$ $13 - 7 = \square$ $9 + \square = 18$

5.

$10 - 2 = \square$ $4 + 7 = \square$ $13 - 4 = \square$

$11 - 2 = \square$ $5 + 7 = \square$ $14 - 5 = \square$

$12 - 2 = \square$ $6 + 7 = \square$ $15 - 6 = \square$

6.

$\square$ twos $= 12$ 3 sixes $= \square$ $\square$ threes $= 15$

$\square$ nines $= 18$ 2 fives $= \square$ $\square$ sevens $= 14$

$\square$ fours $= 16$ 9 twos $= \square$ $\square$ eights $= 16$

PRACTICE

1. $7 + \square = 11$ $8 + \square = 13$ $5 + 6 = \square$

      $7 + \square = 12$ $8 + \square = 14$ $6 + 6 = \square$

      $7 + \square = 13$ $8 + \square = 15$ $7 + 6 = \square$

2. $15 - 9 = \square$ $13 - \square = 7$ $11 - 8 = \square$

      $16 - 9 = \square$ $14 - \square = 7$ $12 - 8 = \square$

      $17 - 9 = \square$ $15 - \square = 7$ $13 - 8 = \square$

3. $3 + 8 = \square + 10$ $4 + 8 = \square + 10$

      $9 + 4 = 10 + \square$ $9 + 7 = 10 + \square$

      $2 + 9 = \square + 10$ $5 + 9 = \square + 10$

4. $18 - 8 = \square$ $12 - \square = 6$ $\square - 9 = 2$

      $16 - 8 = \square$ $12 - \square = 8$ $\square - 9 = 3$

      $14 - 8 = \square$ $12 - \square = 10$ $\square - 9 = 4$

5. $12 - 3 = \square$ $6 + 7 = \square$ $17 - 8 = \square$

      $13 - 5 = \square$ $7 + 8 = \square$ $15 - 7 = \square$

      $14 - 7 = \square$ $8 + 9 = \square$ $13 - 6 = \square$

PRACTICE

1. Which sets match? Write the letters.

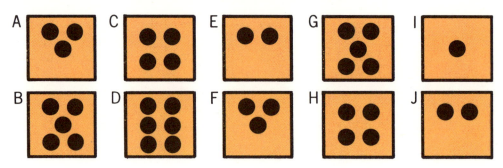

Find the sets. Write the letters.

2. a. Two sets that make 6.

    b. Two sets that make 7.

    c. Two sets that make 8.

    d. Two sets that make 5.

    e. Two sets that make 4.

3. a. A set that has 2 more than another set.

    b. A set that has 2 less than another set.

    c. A set that has 1 more than another set.

    d. A set that has 1 less than another set.

4. a. Three sets that make 6.

    b. Three sets that make 7.

    c. Three sets that make 8.

PRACTICE

1. Add 10 to each number.   Subtract 10 from each.

A 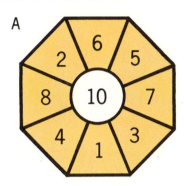   B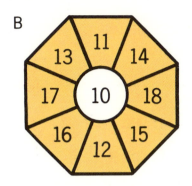

2.
11 − 9 = ☐   13 − 8 = ☐   12 − 4 = ☐

13 − 5 = ☐   12 − 6 = ☐   11 − 6 = ☐

12 − 9 = ☐   13 − 7 = ☐   12 − 7 = ☐

13 − 4 = ☐   11 − 7 = ☐   13 − 9 = ☐

3. What numeral goes in each box?

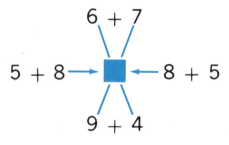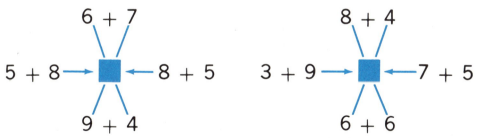

4.   7 + 4 = ☐   6 + 5 = ☐   8 + 3 = ☐

11 − 5 = ☐   11 − 3 = ☐   11 − 4 = ☐

165

1. Write the numerals.

seven	one	eight	ten	five
two	four	three	six	nine

2. 4 tens + 7 ones = ☐   9 tens + 5 ones = ☐
6 tens + 2 ones = ☐   10 tens + 0 ones = ☐
3 tens + 3 ones = ☐   2 tens + 4 ones = ☐

3. Just after:

9?   16?   44?   97?   36?   11?

Just before:

7?   13?   51?   75?   100?   82?

4.
4	2	5	4	3	2	3
3	1	2	2	5	5	3
2	3	3	2	2	3	3
☐	☐	☐	☐	☐	☐	☐

5. What numerals are missing?

5 ___ 15 20 ___ ___ 35 ___ ___ ___

72 74 ___ ___ ___ 82 ___ ___ 88 ___

10 20 ___ 40 ___ ___ 70 ___ 90 ___

166

SOMETHING SPECIAL FOR YOU

1. Joe has 8 picture cards.
   He wants to keep 3 or 4.
   How many can he give away?

2. The first shelf has 7 books on it.
   The second shelf has 10 books.
   A third shelf has more than 7 and
   fewer than 10 books. How many
   books could be on this shelf?

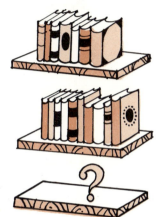

3. Ann has 15¢. She wants to buy
   clothes for her doll. What can
   Ann buy for 15¢?

4. Tom put two nickels and two cents in his
   bank. He took out five cents for ice
   cream. How much is left in his bank?

5. Can Tom buy 2 tops and 2 balloons
   with the money left in his bank?

SOMETHING SPECIAL FOR YOU

1. Write these numerals on your paper: 12, 14, 15.

Find all the names for 12, 14, and 15.

Write the letter for each name under the numerals.

a. 2 sixes + 2

b. 3 fours

c. 3 threes + 6

d. (17 − 2) − 3

e. 2 sevens + 2 − 2

f. (8 + 4) + (5 − 2)

g. (9 − 3) + 6

h. 2 sevens

i. 5 twos + 4

j. 3 fives

k. 2 threes + 3 twos

l. 2 sixes + 1 three

m. (12 − 5) + 7

n. 3 + 4 + (1 five)

o. (10 − 5) + (5 twos)

p. 9 + 1 + 1 + 1

q. 18 − 3

r. 20 − (2 + 2 + 2)

2. Write four different names for 11.

Write four different names for 13.

3. Follow the arrows and add. What is the answer?

REVIEWING FOR THE YEAR

1. 1 dime = ☐ pennies   1 nickel = ☐ pennies
1 dime = ☐ nickels   3 nickels = ☐ pennies
5¢ + 4¢ + 1¢ = ☐ ¢   2¢ + 3¢ + 3¢ = ☐ ¢
5¢ + 5¢ + 5¢ = ☐ ¢   1¢ + 3¢ + 6¢ = ☐ ¢

2. 12 = 10 + ☐   14 = 10 + ☐   19 = 10 + ☐
15 = 10 + ☐   17 = 10 + ☐   13 = 10 + ☐
18 = 10 + ☐   16 = 10 + ☐   11 = 10 + ☐

3. 88 = ☐ tens + △ ones   **4.** 10 = ☐ twos
16 = ☐ ten + △ ones   8 = ☐ fours
54 = ☐ tens + △ ones   6 = ☐ twos
79 = ☐ tens + △ ones   10 = ☐ fives
41 = ☐ tens + △ one   8 = ☐ twos
10 = ☐ ten + △ ones   6 = ☐ threes

5. How many ●'s in all?

A
☐ = △ sixes   ☐ = △ twos
☐ = △ fours   ☐ = △ threes

How many ●'s in all?

B
☐ = △ fives
☐ = △ threes

169

6. Which pictures show fourths?

Which pictures show halves?

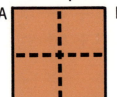 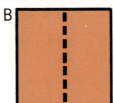 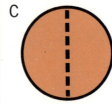 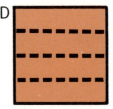

7. What time does each clock show?

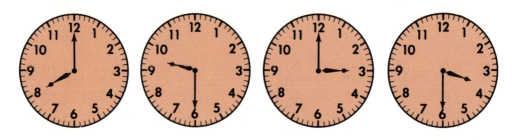

8.   6 +   6 = ☐, so 6 + 7 = ☐.

   7 +   7 = ☐, so 7 + 8 = ☐.

   6 + 10 = ☐, so 6 + 9 = ☐.

   10 +   4 = ☐, so 9 + 4 = ☐.

9. 10 − 5 = ☐, so 10 − 6 = ☐.

   12 − 6 = ☐, so 12 − 7 = ☐.

   14 − 7 = ☐, so 14 − 6 = ☐.

   16 − 8 = ☐, so 16 − 9 = ☐.

INDEX